An Introduction to Laplacian Spectral Distances and Kernels

Theory, Computation, and Applications

Synthesis Lectures on Visual Computing
Computer Graphics, Animation, Computational Photography, and Imaging

Editor
Brian A. Barsky, *University of California, Berkeley*

This series presents lectures on research and development in visual computing for an audience of professional developers, researchers, and advanced students. Topics of interest include computational photography, animation, visualization, special effects, game design, image techniques, computational geometry, modeling, rendering, and others of interest to the visual computing system developer or researcher.

An Introduction to Laplacian Spectral Distances and Kernels: Theory, Computation, and Applications
Giuseppe Patanè

ISBN: 978-3-031-01465-9   paperback
ISBN: 978-3-031-02593-8   ebook

DOI: 10.1007/978-3-031-02593-8

A Publication in the Springer series
Synthesis Lectures on Visual Computing: Computer Graphics, Animation,
Computational Photography, and Imaging

Lecture #29
Series Editor: Brian A. Barsky, *University of California, Berkeley*
Series ISSN
Print 2469-4215   Electronic 2469-4223

An Introduction to Laplacian Spectral Distances and Kernels

Theory, Computation, and Applications

Giuseppe Patanè
CNR-IMATI

SYNTHESIS LECTURES ON VISUAL COMPUTING: COMPUTER GRAPHICS, ANIMATION, COMPUTATIONAL PHOTOGRAPHY, AND IMAGING #29

ABSTRACT

In geometry processing and shape analysis, several applications have been addressed through the properties of the Laplacian spectral kernels and distances, such as commute-time, biharmonic, diffusion, and wave distances.

Within this context, this book is intended to provide a common background on the definition and computation of the Laplacian spectral kernels and distances for geometry processing and shape analysis. To this end, we define a unified representation of the isotropic and anisotropic discrete Laplacian operator on surfaces and volumes; then, we introduce the associated differential equations, i.e., the harmonic equation, the Laplacian eigenproblem, and the heat equation. Filtering the Laplacian spectrum, we introduce the Laplacian spectral distances, which generalize the commute-time, biharmonic, diffusion, and wave distances, and their discretization in terms of the Laplacian spectrum. As main applications, we discuss the design of smooth functions and the Laplacian smoothing of noisy scalar functions.

All the reviewed numerical schemes are discussed and compared in terms of robustness, approximation accuracy, and computational cost, thus supporting the reader in the selection of the most appropriate with respect to shape representation, computational resources, and target application.

KEYWORDS

Laplace-Beltrami operator, Laplacian spectrum, harmonic equation, Laplacian eigenproblem, heat equation, diffusion geometry, Laplacian spectral distance and kernels, spectral geometry processing, shape analysis, numerical analysis

To my family

Contents

List of Figures

List of Tables

Preface

In geometry processing and shape analysis, several applications have been addressed through the properties of the Laplacian spectral kernels and distances, such as commute-time, biharmonic, diffusion, and wave distances. Spectral distances are easily defined through a filtering of the Laplacian eigenpairs and include random walks [FPS05, RS13], heat diffusion [BBK+10, BBOG11, CL06, GBAL09, LKC06, LSW09], biharmonic [LRF10, Rus11b], and wave kernel [BB11a, ASC11] distances. Biharmonic [LRF10, Rus11b] and diffusion [BBK+10, BBOG11, CL06, GBAL09, LKC06, LSW09, PS13b] distances provide a trade-off between a nearly geodesic behavior for small distances and the encoding of global surface properties for large distances, thus guaranteeing an intrinsic and multi-scale characterization of the input shape. The heat kernel [BBG94] is also central in diffusion geometry [BN03, CL06, GK06, Sin06], dimensionality reduction with spectral embeddings [BN03, XHW10], and data classification [SK03].

As main applications, we mention the multi-scale approximation of functions [PF10] and gradients [LSW09], shape segmentation and comparison through heat kernel shape descriptors, auto-diffusion functions, and diffusion distances. Laplacian spectral distances have been applied to shape segmentation [dGGV08] and comparison [BBOG11, GBAL09, Mem09, OMMG10, SOG09] with multi-scale and isometry-invariant signatures [DRW10, LKC06, MS05, Mem11, RBBK10, Rus07, MS09]. In fact, they are intrinsic to the input shape, invariant to isometries, multi-scale, and robust to noise and tessellation. Additional applications include image smoothing [ZH08], geometric characterizations [EH09] and embeddings [LWH03] of graphs, and shape analysis [BMR+16, CRA+16, RCB+16]. The diffusion kernel and distance also play a central role in several applications, such as dimensionality reduction with spectral embeddings [BN03, XHW10]; data visualization [BN03, HAvL05, RS00, TSL00], representation [CWS03, SK03, ZGL03], and classification [NJW01, SM00, ST07].

Our book is intended to provide a common background on the definition and computation of the Laplacian spectral kernels and distances for geometry processing and shape analysis. All the reviewed numerical schemes are discussed and compared in terms of robustness, approximation accuracy, and computational cost, thus supporting the reader in the selection of the most appropriate with respect to shape representation, computational resources, and target application. Indeed, this book is complementary to previous work, which has been focused mainly on specific applications, such as mesh filtering [Tau99], surface coding and spectral partitioning [KG00], 3D shape deformation based on differential coordinates [Sor06], spectral methods [ZvKD07, Pat16b], and Laplacian eigenfunctions [Lev06] for geometry processing and diffusion shape analysis [BCA12].

First, we define a unified representation of the isotropic and anisotropic discrete Laplacian operator on surfaces and volumes (Chapter 1); then, we introduce the associated differential equations. For the harmonic equation and the Laplacian eigenproblem, we focus on the stability and accuracy of numerical solvers, also presenting their main applications. This discussion provides the background for a detailed analysis of the heat equation (Chapter 2) and allows us to identify the main limitations (e.g., computational cost, storage overhead, selection of user-defined parameters) of previous work on the approximation of the diffusion distances, which is based mainly on the evaluation of the Laplacian spectrum and on linear approximations of the exponential matrix. For the heat equation, we discuss the selection of the time scale and the main approaches for the computation of the solution to the heat equation, such as linear, polynomial, and rational approximations.

Filtering the Laplacian spectrum, we introduce the Laplacian spectral distances (Chapter 3), which generalize the commute-time, biharmonic, diffusion and wave distances, and their discretization in terms of the Laplacian spectrum. The growing interest on these distances is motivated by their capability of encoding local geometric properties (e.g., Gaussian curvature, geodesic distance) of the input shape, their intrinsic and multi-scale definition with respect to the input shape, their invariance to isometries, shape-awareness, robustness to noise, and tessellation. While previous work has been focused mainly on surfaces discretized as triangle meshes, we introduce a unified representation of the spectral distances and kernels, which is independent of the selected Laplacian weights, of the surface or volume representation as polygonal mesh, point set, tetrahedral or voxel grid. From this general representation, we show that the characteristic properties of the spectral distances are guided mainly by the filter that is applied to the Laplacian eigenpairs.

The expensive cost for the computation of the Laplacian spectrum and the sensitiveness of multiple Laplacian eigenvalues to surface discretization generally preclude an accurate evaluation of the spectral kernels and distances on large data sets. To discuss these problems, we review and compare different methods for the numerical evaluation of the spectral distances and kernels (Chapter 4). In particular, we detail their spectrum-free computation, which is defined through a polynomial or rational approximation of the filter function. The resulting computational scheme only requires the solution of sparse linear systems, is not affected by the Gibbs phenomenon, and is independent of the representation of the input domain, the selected Laplacian weights, and the evaluation of the Laplacian spectrum.

As main applications (Chapter 5), we will discuss the design of smooth functions, whose maxima, minima, and saddles are selected by the user or imported from a template function, and the Laplacian smoothing of noisy scalar functions, without and with constraints on the preservation of their critical points. Finally (Chapter 6), we conclude our review with a discussion of open questions and challenges.

Giuseppe Patanè
June 2017

Acknowledgments

Shapes are courtesy of AIM@SHAPE Repository, and tet-meshes were generated by the TET-GEN software (`http://wias-berlin.de/software/tetgen/`).

Giuseppe Patanè
June 2017

CHAPTER 1

Laplace-Beltrami Operator

We review the isotropic and anisotropic Laplace-Beltrami operator and introduce a unified representation of the corresponding Laplacian matrix for surfaces and volumes. Additional results have been presented in [Sor06, Tau99, KG00, ZvKD07].

Let $\mathcal{N}$ be a smooth compact surface (with or without boundary), possibly with boundary, equipped with a Riemannian metric and let us consider the *scalar product* $\langle f, g \rangle_2 := \int_{\mathcal{N}} f(\mathbf{p})g(\mathbf{p})d\mathbf{p}$ defined on the space $\mathcal{L}_2(\mathcal{N})$ of square integrable functions on $\mathcal{N}$ and the corresponding norm $\| \cdot \|_2$. Then, the *intrinsic smooth Laplace-Beltrami operator* $\Delta := -\mathrm{div}(\mathrm{grad})$ satisfies the following properties [Ros97]:

- *self-adjointness*: $\langle \Delta f, g \rangle_2 = \langle f, \Delta g \rangle_2, \forall f, g$;

- *positive semi-definiteness*: $\langle \Delta f, f \rangle_2 \geq 0, \forall f$. In particular, the Laplacian eigenvalues are positive;

- *null eigenvalue*: the smallest Laplacian eigenvalue is null and the corresponding eigenfunction ϕ, $\Delta \phi = 0$, is constant;

- *locality*: the value $\Delta f(\mathbf{p})$ does not depend on $f(\mathbf{q})$, for any couple of distinct points $\mathbf{p}, \mathbf{q}$; and

- *linear precision*: if $\mathcal{N}$ is planar and f is linear, then $\Delta f = 0$.

The *anisotropic Laplace-Beltrami operator* [ARAC14] is defined as $\Delta_{\mathbf{D}} f = \mathrm{div}(\mathbf{D}\nabla f)$, where $\mathbf{D}$ is a 2×2 matrix applied to vectors belonging to the tangent plane and controls the direction and strength of the deviation from the isotropic case. The tensor $\mathbf{D} := \mathrm{diag}(\varphi_\alpha(\kappa_m), \varphi_\alpha(\kappa_M))$ takes into account the directions and the values κ_m, κ_M of low and high curvature, where the filter is $\varphi_\alpha(s) := (1 + \alpha|s|)^{-1}, \alpha > 0$. As $\alpha \to 0$, we get the isotropic Laplace-Beltrami operator (i.e., $\mathbf{D} := \mathbf{I}$). The alternative definition [KTT13] of the anisotropic Laplace-Beltrami operator applies a non-linear factor $\mathbf{D}(\mathbf{v})$, which modifies the magnitude of $\mathbf{D}(\mathbf{v})$ without changing its direction.

We now introduce a unified representation of the Laplacian matrix on surfaces and volumes, which is independent of the underlying discretization. More precisely, we will discuss the discretization of the Laplace-Beltrami operator (Sec. 1.1) on graphs, meshes, and volumes. Then, we will discuss the properties and solution to the harmonic equation (Sec. 1.2) and to the Laplacian eigenproblem (Sec. 1.3).

In the book examples, the level sets of a given function, kernel, or distance are associated with iso-values uniformly sampled in its range.

1.1 DISCRETE LAPLACIANS AND SPECTRAL PROPERTIES

Let us consider a (triangular, polygonal, volumetric) mesh $\mathcal{M} := (\mathcal{P}, T)$, which discretizes a domain $\mathcal{N}$, where $\mathcal{P} := \{\mathbf{p}_i\}_{i=1}^n$ is the set of n vertices and T is the connectivity graph (Fig. 1.1). On $\mathcal{M}$, a piecewise linear scalar function $f : \mathcal{M} \rightarrow \mathbb{R}$ is defined by linearly interpolating the values $\mathbf{f} := (f(\mathbf{p}_i))_{i=1}^n$ of f at the vertices using barycentric coordinates. For point sets, f is defined only at $\mathcal{P}$ and T is the k-nearest neighbor graph.

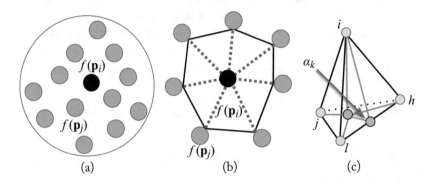

(a) (b) (c)

Figure 1.1: Neighbor and Laplacian stencil for a (a) point set, (b) triangle, and (c) tetrahedral mesh.

We represent the Laplace-Beltrami operator on surface and volume meshes in a unified way as $\tilde{\mathbf{L}} := \mathbf{B}^{-1}\mathbf{L}$, where $\mathbf{B}$ is a sparse, symmetric, positive definite matrix (*mass matrix*) and $\mathbf{L}$ is sparse, symmetric, and positive semi-definite (*stiffness matrix*). We also assume that the entries of $\mathbf{B}$ are positive and that the sum of each row of $\mathbf{L}$ is null. In particular, we consider the $\mathbf{B}$-scalar product $\langle \mathbf{f}, \mathbf{g} \rangle_{\mathbf{B}} := \mathbf{f}^\top \mathbf{B} \mathbf{g}$ and the induced norm $\|\mathbf{f}\|_{\mathbf{B}}^2 := \mathbf{f}^\top \mathbf{B} \mathbf{f}$. Analogously to the continuous case, the Laplacian matrix satisfies the following properties:

- *self-adjointness*: $\tilde{\mathbf{L}}$ is adjoint with respect to the $\mathbf{B}$-scalar product; i.e., $\langle \tilde{\mathbf{L}}\mathbf{f}, \mathbf{g} \rangle_{\mathbf{B}} = \langle \mathbf{f}, \tilde{\mathbf{L}}\mathbf{g} \rangle_{\mathbf{B}} = \mathbf{f}^\top \mathbf{L} \mathbf{g}$. If $\mathbf{B} := \mathbf{I}$, then this property reduces to the symmetry of $\mathbf{L}$;

- *positive semi-definiteness*: $\langle \tilde{\mathbf{L}}\mathbf{f}, \mathbf{f} \rangle_{\mathbf{B}} = \mathbf{f}^\top \mathbf{L} \mathbf{f} \geq 0$. In particular (Sec. 1.3.1), the Laplacian eigenvalues are positive;

- *null eigenvalue*: by construction, we have that $\tilde{\mathbf{L}}\mathbf{1} = \mathbf{0}$; and

- *locality*: since the weight $w(i, j)$ is not null for each edge (i, j), the value $(\tilde{\mathbf{L}}\mathbf{f})_i$ depends only on the f-values at $\mathbf{p}_i$ and its 1-star neighbor $\mathcal{N}(i) := \{j : (i, j) \text{ edge}\}$.

For a detailed discussion of these properties with respect to the selected Laplacian weights, we refer the reader to [WMKG07]. We now discuss the discretization of the Laplace-Beltrami operator on graphs, meshes, volumes (Sec. 1.1.1), and point sets (Sec. 1.1.2).

1.1.1   LAPLACIAN ON GRAPHS, MESHES, AND VOLUMES

Associating a set $\{w(i, j)\}_{i,j}$ of positive weights with the edges (i, j) of T, the entries of the stiffness matrix are defined as $L(i, j) = -\sum_{k \neq i} w(i, k) + w(i, j)$. The entries of the mass matrix $\mathbf{B}$ are normalization coefficients that take into account the geometry of the input domain.

On *graphs* [Chu97], the weights of the stiffness matrix are equal to 1 for each edge and zero otherwise; each diagonal entry of the mass matrix is equal to the valence of the corresponding node. On *triangle meshes*, the *stiffness matrix* $\mathbf{L}$ and the *mass matrix* $\mathbf{B}$ of the linear FEM Laplacian weights [RWP06, VL08] are defined as

$$L(i, j) := \begin{cases} w(i, j) := -\frac{\cot \alpha_{ij} + \cot \beta_{ij}}{2} & j \in \mathcal{N}(i), \\ -\sum_{k \in \mathcal{N}(i)} w(i, k) & i = j, \end{cases}$$

$$B(i, j) := \begin{cases} \frac{|t_r| + |t_s|}{12} & j \in \mathcal{N}(i), \\ \frac{\sum_{k \in \mathcal{N}(i)} |t_k|}{6} & i = j, \end{cases}$$

where $\mathcal{N}(i)$ is the 1-star of the vertex i; α_{ij}, β_{ij} are the angles opposite to the edge (i, j) (Fig. 1.1b); t_r, t_s are the triangles that share the edge (i, j); and $|t|$ is the area of the triangle t. Lumping the mass matrix $\mathbf{B}$ to the diagonal matrix $\mathbf{D}$, $D(i, i) = \frac{1}{3} \sum_{t \in \mathcal{N}(i)} |t|$, whose entries are the areas of the Voronoi regions, $\tilde{\mathbf{L}}$ reduces to the Laplacian matrix $\mathbf{D}^{-1}\mathbf{L}$ with *Voronoi-cotangent weights* [DMSB99], which extend the cotangent weights introduced in [PP93] ($\mathbf{B} := \mathbf{I}$). The *mean-value weights* [Flo03] have been derived from the mean value theorem for harmonic functions and are always positive. In [CLB$^+$09], the weak formulation of the Laplacian eigenproblem is achieved by selecting a set of volumetric test functions, which are defined as $k \times k \times k$ B-splines (e.g., $k := 4$) and restricted to the input shape. For the *anisotropic Laplacian operator* [ARAC14], the entries of $\mathbf{L}$ are a variant of the cotangent weights (i.e., with respect to different angles) and the entries of the diagonal mass matrix $\mathbf{B}$ are the areas of the Voronoi regions.

While the Laplace-Beltrami operator depends only on the Reimannian metric (*intrinsic property*), its discretization is generally affected by the quality of the input triangulation [She02, HPW06]. For instance, two (simplicial) isometric surfaces with two different triangulations are associated with two different Laplacian matrices. According to [BS07], the cotangent weights are non-negative if and only if the input triangulation is Delaunay and the corresponding Laplacian matrix is more accurate than the one evaluated on the original mesh.

We briefly recall [DZM07, LXH15, LXFH15] that a triangulation of a piecewise flat surface is a Delaunay triangulation if and only if all its interior edges are locally Delaunay (i.e., the sum of the angles opposite to an edge in the adjacent triangles does not exceed π). Furthermore, the minimum of the Dirichlet energy of a piecewise linear function, on all the possible triangulations of a piecewise flat surface $\mathcal{M}$, is attained at the Delaunay triangulation of $\mathcal{M}$ and the corresponding discrete Laplace-Beltrami operator is intrinsic to the input surface.

On *polygonal meshes*, the Laplacian discretization in [AW11, HKA15] generalizes the Laplacian matrix with cotangent weights to surface meshes with non-planar, non-convex faces. An approximation of the Laplace-Beltrami operator with point-wise convergence has been proposed in [BSW08].

On *volumes* represented as a tetrahedral meshes [ACSYD05, LTDZ09, TLHD03], the entries of the stiffness matrix are (Fig. 1.1c) $L(i,j) := w(i,j) := \frac{1}{6}\sum_{k=1}^{n} l_k \cot \alpha_k$ for each edge (i,j), $L(i,i) := -\sum_{j \in \mathcal{N}(i)} w(i,j)$, and zero otherwise; the diagonal mass matrix $\mathbf{B}$ encodes the tetrahedral volume at each vertex.

1.1.2   LAPLACIAN MATRIX OF POINT SETS

In [BN03, BN06, BN08, BSW09], the Laplace-Beltrami operator on a point set $\mathcal{P}$ has been discretized as the Laplacian matrix

$$L(i,j) := \frac{1}{nt(4\pi t)^{3/2}} \begin{cases} \exp\left(-\frac{\|\mathbf{p}_i - \mathbf{p}_j\|_2}{4t}\right) & i \neq j, \\ -\sum_{k \neq i} L(i,k) & i = j. \end{cases}$$

To guarantee the sparsity of the Laplacian matrix, for each point $\mathbf{p}_i$ we consider only the entries $L(i,j)$ related to the points $\{\mathbf{p}_j\}_{j \in \mathcal{N}_{p_i}}$ that are closest to $\mathbf{p}_i$ with respect to the Euclidean distance. In this case, we select either the k-nearest neighbor or the points that belong to a sphere centerd at $\mathbf{p}_i$ and with radius σ (Fig. 1.1a). As described in [DS05, MN03], the choice of σ can be adapted to the local sampling density $\epsilon := k(\pi\sigma^2)^{-1}$ and the curvature of the surface underlying $\mathcal{P}$. The computation of the k- or σ-nearest neighbor graph takes $\mathcal{O}(n \log n)$-time [AMN+98, Ben75], where n is the number of input points.

Starting from this approach, a new discretization [LPG12] has been achieved through a finer approximation of the local geometry of the surface at each point through its Voronoi cell. More precisely, as $t \to 0$ the stiffness and mass matrix are defined as

$$L(i,j) := \begin{cases} \frac{1}{4\pi t^2} \exp\left(-\frac{\|\mathbf{p}_i - \mathbf{p}_j\|_2^2}{4t}\right) & i \neq j, \\ -\sum_{k \neq i} L(i,k) & i = j, \end{cases} \qquad B(i,i) = v_i,$$

and v_i is the area of the Voronoi cell associated with the point $\mathbf{p}_i$. The Voronoi cell of $\mathbf{p}_i$ is approximated by projecting the points of a neighbor of $\mathbf{p}_i$ on the estimated tangent plane to $\mathcal{M}$ at $\mathbf{p}_i$. If $\mathbf{B} := \mathbf{I}$, then this approximation reduces to the previous one and both approaches converge to the Laplace-Beltrami operator, as $t \to 0^+$.

1.2 HARMONIC EQUATION

The *harmonic function* $h : \mathcal{N} \to \mathbb{R}$ is the solution of the Laplace equation $\Delta h = 0$ with Dirichlet boundary conditions $h|_{\mathcal{S}} = h_0$, $\mathcal{S} \subset \mathcal{N}$. We recall that a harmonic function

- minimizes the *Dirichlet energy* $\mathcal{E}(h) := \int_{\mathcal{N}} \|\nabla h(\mathbf{p})\|_2^2 \mathrm{d}\mathbf{p}$;

- satisfies the *locality property*, i.e., if $\mathbf{p}$ and $\mathbf{q}$ are two distinct points, then $\Delta h(\mathbf{p})$ is not affected by the value of h at $\mathbf{q}$; and

- verifies the relation

$$h(\mathbf{p}) = (2\pi R)^{-1} \int_{\Gamma} h(s)\mathrm{d}s = (\pi R^2)^{-1} \int_{\mathcal{B}} h(\mathbf{q})\mathrm{d}\mathbf{q},$$

  where $\mathcal{B} \subseteq \mathcal{N}$ is a disc of center $\mathbf{p}$, radius R, and boundary Γ (*mean-value theorem*).

According to the *maximum principle* [Ros97], a harmonic function has no local extrema other than at constrained vertices. If all constrained minima are assigned the same global minimum value and all constrained maxima are assigned the same global maximum value, then all the constraints will be extrema in the resulting field. Harmonic and poly-harmonic (i.e., $\Delta^i h = 0$) functions have been applied to volumetric parameterization [LGW+07, LXW+10], to the definition of shape descriptors with pairs of surface points [ZTZX13] and coupled biharmonic bases [KBB+13], to shape approximation [FW12], and deformation [JMD+07, JBPS14, WPG12].

The harmonic equation is approximated at the vertices of $\mathcal{M}$ as the homogeneous linear system $\mathbf{Lf} = \mathbf{0}$, with initial conditions $f(\mathbf{p}_i) = a_i$, $i \in \mathcal{I} \subseteq \{1, \ldots, n\}$. According to the *Euler formula* $\chi(\mathcal{M}) = m - s + M$ (Sec. 5.2.1), the number of minima m, maxima M, and saddles s of a harmonic function depends on the Dirichlet boundary conditions, which determine the maxima and minima of the resulting harmonic function. In particular, a harmonic function with one maximum and one minimum has a minimal number of $2g$ saddles, where g is the genus of $\mathcal{M}$ (Fig. 1.2). Harmonic functions are efficiently computed in $\mathcal{O}(n)$ time with iterative solvers of sparse linear systems; their computation is stable for the mean-value weights while negative Voronoi cotangent weights generally induce local undulations in the resulting harmonic function. Main applications include surface quadrangulation [DKG05, NGH04], the definition of volumetric mappings [LGQ09, LXW+10, LI14, MCK08, MC10, XYGL13], and biharmonic distances [OBCS+12, LRF10, Rus11b] (Sec. 3.4.3).

1.3 LAPLACIAN EIGENPROBLEM

Since the Laplace-Beltrami operator is self-adjoint and positive semi-definite, it has an *orthonormal eigensystem* $\mathcal{B} := \{(\lambda_n, \phi_n)\}_{n=0}^{+\infty}$, $\Delta \phi_n = \lambda_n \phi_n$, in $\mathcal{L}_2(\mathcal{N})$. In the following, we assume that

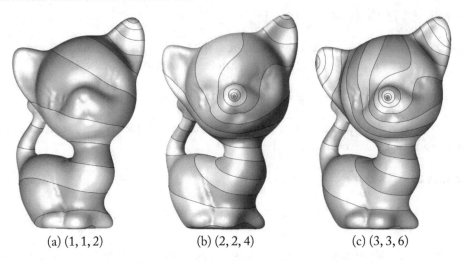

(a) $(1, 1, 2)$ (b) $(2, 2, 4)$ (c) $(3, 3, 6)$

Figure 1.2: Level sets and critical points (m, M, s) of harmonic functions with (a) two, (b) four, and (c) six Dirichlet boundary conditions. The insertion of new initial constraints locally affects the resulting harmonic function.

the Laplacian eigenvalues are increasingly ordered; in particular, $\lambda_0 = 0$ and the corresponding eigenfunction is the constant function $\phi_0 = 1$. Using the orthonormality and completeness of the Laplacian eigenfunctions in $\mathcal{L}_2(\mathcal{N})$, any function can be represented as a linear combination of the eigenfunctions as $f = \sum_{n=0}^{+\infty} \langle f, \phi_n \rangle_2 \phi_n$, where $\langle f, \phi_n \rangle_2 \phi_n$ is the projection of f on ϕ_n. Furthermore, the function Δf is expressed in terms of the Laplacian spectrum as $(\Delta f)(\mathbf{p}) = \sum_{n=0}^{+\infty} \lambda_n \langle f, \phi_n \rangle_2 \phi_n(\mathbf{p})$ (*spectral decomposition theorem*). A deeper discussion of the analogies between the heat kernel, the Fourier analysis, and wavelets has been presented in [HVG11, BEKB15].

The Laplacian eigenfunctions are intrinsic to the input shape and those ones related to larger eigenvalues correspond to smooth and slowly-varying functions. Increasing the eigenvalues, the corresponding eigenfunctions generally show rapid oscillations (Fig. 1.3). From the Laplacian spectrum, we can estimate geometric and topological properties of the input shape. For instance, we can compute the surface area, as the sum of the Laplacian eigenvalues; estimate the Euler characteristic of a surface with genus $g \geq 2$ through the relation [Nad88] $m_j \leq 2j - 2\chi(\mathcal{M}) + 3$, where m_j is the multiplicity of λ_j; and evaluate the total Gaussian curvature [RWP06]. If two shapes are isometric, then they have the same Laplacian spectrum (*isospectral property*); however, the viceversa does not hold [GS02, ZGLG12] and we cannot recover the metric of a given surface. In the following, we introduce the discretization (Sec. 1.3.1) and stability of the computation (Sec. 1.3.2) of the Laplacian eigenpairs.

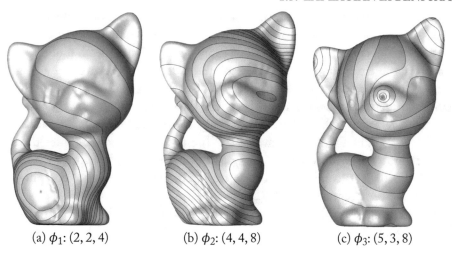

(a) ϕ_1: $(2, 2, 4)$ (b) ϕ_2: $(4, 4, 8)$ (c) ϕ_3: $(5, 3, 8)$

Figure 1.3: Level sets and number of critical points (maxima, minima, saddles) of different Laplacian eigenfunctions (linear FEM weights).

1.3.1 DISCRETE LAPLACIAN EIGENFUNCTIONS

To introduce the discrete Laplacian eigenpairs, the Laplacian eigenproblem is converted to its weak formulation $\langle \Delta\phi, \psi \rangle_2 = \lambda \langle \phi, \psi \rangle_2$ [All07], where ψ is a test function. The weak formulation is then discretized as $\mathbf{L}\mathbf{x} = \lambda \mathbf{B}\mathbf{x}$. Here, $\mathbf{L}$, $L(i, j) := \langle \Delta\psi_i, \psi_j \rangle_2$, is the stiffness matrix and $\mathbf{B}$, $B(i, j) := \langle \psi_i, \psi_j \rangle_2$, is the mass matrix. The *generalized Laplacian eigensystem* $\{(\lambda_i, \mathbf{x}_i)\}_{i=1}^n$ satisfies the identity $\mathbf{L}\mathbf{x}_i = \lambda_i \mathbf{B}\mathbf{x}_i$ and the eigenvectors are orthonormal with respect to the $\mathbf{B}$-scalar product, i.e., $\langle \mathbf{x}_i, \mathbf{x}_j \rangle_\mathbf{B} = \mathbf{x}_i^\top \mathbf{B}\mathbf{x}_j = \delta_{ij}$. In particular, $\lambda_0 = 0$ and the corresponding eigenvector is $\mathbf{1}$, i.e., the vector whose entries are equal to 1 (in fact, the sum $\mathbf{L}\mathbf{1}$ of the rows of $\mathbf{L}$ is null). In the discrete setting, the spectral decomposition theorem becomes $\tilde{\mathbf{L}}\mathbf{f} = \sum_{i=1}^n \lambda_i \langle \mathbf{f}, \mathbf{x}_i \rangle_\mathbf{B} \mathbf{x}_i = \mathbf{X}\Lambda\mathbf{X}^\top \mathbf{B}\mathbf{f}$, where $\mathbf{X}$ is the eigenvectors' matrix and Λ is the diagonal matrix of the eigenvalues. The discrete Laplacian eigenfunctions generally have a global support (i.e., they are null only at some isolated points) and eigenfunctions with a compact support can be calculated by minimizing the corresponding ℓ_1 norm [NVT+14]. Indeed, the *generalized spectral decomposition* of the Laplacian matrix is characterized by the relations

$$\mathbf{L} = \mathbf{X}\Lambda\mathbf{X}^\top \mathbf{B}, \qquad \mathbf{X}^\top \mathbf{B}\mathbf{X} = \mathbf{I}. \tag{1.1}$$

For the computation of the Laplacian eigenvectors, numerical methods generally exploit the sparsity of the Laplacian matrix and reduce the high-dimensional eigenproblem to one of lower dimension, by applying a coarsening step. The solution is efficiently calculated in the low-dimensional space and then mapped back to the initial dimension through a refinement step. Main examples include the algebraic multi-grid method [Fal06], Arnoldi iterations [LS96, Sor92], and the Nystrom method [FBCM04]. Even though the eigenvalues and eigenvectors are computed in super-linear time [VL08], this computational cost and the required $\mathcal{O}(n^2)$ storage are expensive for densely sampled domains. Indeed, modifications of the Laplacian eigenproblem are applied to locally compute specific sub-parts of the Laplacian spectrum. For instance, the shift method evaluates the spectrum $(\lambda_i - \lambda, \mathbf{x}_i)_{i=1}^n$ of $(\tilde{\mathbf{L}} - \lambda\mathbf{I})$ and then provides the eigenpairs associated with a spectral band centerd around a value λ. To compute the larger eigenvalue and the corresponding eigenvector, the inverse method considers the spectrum $(\lambda_i^{-1}, \mathbf{x}_i)_{i=2}^n$ of the pseudo-inverse $\tilde{\mathbf{L}}^\dagger$. The power method computes the eigenpairs $(\lambda_i^k, \mathbf{x}_i)_{i=2}^n$ of the sequence of matrices $(\tilde{\mathbf{L}}^k)_{k \geq 1}$ and controls the convergence speed through the selection of k. Finally, pre-conditioners of the Laplacian matrix tailored to computer graphics' applications have been proposed in [KFS13].

In the following, we address the discretization and properties of the Laplacian eigenpairs on surfaces and volumes.

Laplacian Eigenfunctions on Surfaces

In *spectral graph theory*, the Laplacian eigenvectors have been applied to graph partitioning [Fie73, MP93, Kor03] into sub-graphs, which are handled in parallel [AKY99], to graph/mesh layout [DPS02, Kor03], to the reduction of the bandwidth of sparse matrices [BPS93]. In *machine learning*, the Laplacian spectrum have been used for clustering [SS02, Sec. 14] and dimensionality reduction [BN03, XHW10] with spectral embeddings. For instance, a common way to measure the dissimilarity between two graphs is to compute the corresponding spectral decomposition in their own [LD08] or joint [Ume88, CK04] eigenspaces. Figure 1.3 shows the level sets and critical points of a family of Laplacian eigenfunctions (Sec. 5.2.1); here, the level sets correspond to iso-values uniformly sampled in the range of the Laplacian eigenfunctions.

In *geometry processing*, the spectral properties of the uniform discrete Laplacian have been used to design low-pass filters [Tau95]. Successively, this formulation has been refined to include the local geometry of the input surface [DMSB99, KR05, PP93] and it has been applied to implicit mesh fairing [DMSB99, KR05, ZF03] and to fairing functionals [KCVS98, Mal89], which optimize the triangles' shape and/or the surface smoothness [NISA06]. Further applications include mesh watermarking [OTMM01, OMT02], geometry compression [KG00, SCOT03], the computation of the gradient [LSW09], and the multi-scale approximation of functions [PF09, Pat13, PS13a, PF09]. The Laplacian eigenvectors have been used also for embedding a surface of arbitrary genus into the plane [ZSGS04, ZKK02] and mapping a closed genus zero surface into a spherical domain [GGS03].

In *shape analysis*, the Laplacian spectrum has been applied to shape [HDL16, LZ07, ZL05] segmentation, comparison [YLL12], and analysis through nodal domains [RBG$^+$09], correspondence [JZ07, JZvK07], and comparison [MPSF11, RWP06, JZ07]. Mesh Laplacian operators are also associated with a set of differential coordinates for surface deformation [SLCO+04] and quadrangulation with Laplacian eigenfunctions [DKG05]. As detailed in Chapter 3, the Laplacian spectrum is also fundamental to define random walks [RS13], commute-time [BB11a], biharmonic [OBCS+12, Rus11b], wave kernel [BB11a, ASC11], and diffusion distances [BBK+10, BBOG11, CL06, GBAL09, LKC06, LSW09, PS13b].

Laplacian Eigenfunctions on Volumes
Laplacian eigenfunctions on a discrete volumetric domain $\mathcal{M}$ are computed either by diagonalising the corresponding Laplacian matrix (Sec. 1.1.2) or by extending the values of the eigenfunctions computed on the boundary of $\mathcal{M}$ to its interior with barycentric coordinates or non-linear methods (e.g., moving least-squares, radial basis functions) [PF09, PS12]. The computational cost, which is generally high in case of volumetric meshes, is effectively reduced but associated with a lower approximation accuracy. Volumetric Laplacian eigenfunctions have been applied to shape retrieval [JZ07] and to the definition of volumetric [Rus11a] shape descriptors.

1.3.2 STABILITY OF THE LAPLACIAN SPECTRUM

Theoretical results on the sensitivity of the Laplacian spectrum against geometry changes, irregular sampling density and connectivity have been presented in [HPW06, Xu07]. Here, we briefly recall that the instability of the computation of the Laplacian eigenpairs is generally due to repeated or close eigenvalues, with respect to the numerical accuracy of the solver of the eigen-equation. While repeated eigenvalues are quite rare and typically associated with symmetric shapes, numerically close or switched eigenvalues can be present in the spectrum and in spite of the regularity of the input discrete surface. The following discussion will be generalized and applied to the analysis of the stability of the spectrum of the Laplacian spectral operator (Sec. 3.5); it will be useful also for the characterization of the filter function that induces the spectral distances (Sec. 4.3).

To show that the computation of single eigenvalues is numerically stable, we perturb the Laplacian matrix $\tilde{\mathbf{L}}$ by $\epsilon \mathbf{E}$, $\epsilon \to 0$, and compute the eigenpair $(\lambda(\epsilon), \mathbf{x}(\epsilon))$ of the new problem $(\mathbf{B}^{-1}\mathbf{L} + \epsilon \mathbf{E})\mathbf{x}(\epsilon) = \lambda(\epsilon)\mathbf{x}(\epsilon)$, with initial conditions $\mathbf{x}(0) = \mathbf{x}$, $\lambda(0) = \lambda$. The size of the derivative of $\lambda(\epsilon)$ indicates the variation that it undergoes when the matrix $\tilde{\mathbf{L}}$ is perturbed in the direction $(\mathbf{E}, \epsilon)$. By differentiating the previous equation and evaluating the result at $\epsilon = 0$, we obtain that $\mathbf{B}\mathbf{E}\mathbf{x} + \mathbf{L}\mathbf{x}'(0) = \lambda'(0)\mathbf{B}\mathbf{x} + \lambda\mathbf{B}\mathbf{x}'(0)$. Multiplying both sides of this last relation with $\mathbf{x}^\top$, the perturbed eigenvalue $|\lambda'(0)| = |\mathbf{x}^\top \mathbf{B}\mathbf{E}\mathbf{x}| \leq \|\mathbf{E}\mathbf{x}\|_{\mathbf{B}}$ is bounded by the $\mathbf{B}$-norm of $\mathbf{E}\mathbf{x}$. Indeed, the computation of the Laplacian eigenvalue with multiplicity one is stable.

Assuming that λ_k is an eigenvalue with multiplicity m_k and rewriting the characteristic polynomial as $p_{\bar{\mathbf{L}}}(\lambda) = (\lambda - \lambda_k)^{m_k} q(\lambda)$, where $q(\cdot)$ is a polynomial of degree $n - m_k$ and $q(\lambda_k) \neq 0$, we get that $(\lambda - \lambda_k)^{m_k} = \mathcal{O}(\epsilon)/q(\lambda)$; i.e., $\lambda = \lambda_k + \mathcal{O}(\epsilon^{\frac{1}{m_k}})$. It follows that a perturbation $\epsilon := 10^{-m_k}$ produces a change of order 0.1 in λ_k and this amplification becomes more and more evident while increasing the multiplicity of the eigenvalue. According to [GV89, Sec. 7], repeated eigenvalues are generally associated with a numerical instability in the computation of the corresponding eigenvectors; in fact, the ℓ_2-norm of the difference between the generalized eigenvectors $\mathbf{x}_i$, $\mathbf{x}_j$ related to the eigenvalues λ_i, λ_j is bounded as

$$\|\mathbf{x}_i - \mathbf{x}_j\|_2 \leq \epsilon \sum_{j \neq i} \left| \frac{\mathbf{x}_i^\top \mathbf{E} \mathbf{x}_j}{\lambda_i - \lambda_j} \right| + \mathcal{O}(\epsilon^2).$$

Indeed, the computation of the eigenvectors related to multiple or close Laplacian eigenvalues might be unstable. Finally, the Laplacian eigenvalues might be locally switched (i.e., we are not able to numerically distinguish two consecutive eigenvalues) and this situation happens independently of the quality of the discretized surface in terms of point density, angles, and connectivity. This analysis of the stability of the Laplacian spectrum will be generalized to the study of the spectrum of operators induced by properly filtering the Laplace-Beltrami operator (Sec. 3.2.2).

CHAPTER 2

Heat and Wave Equations

We introduce the heat (Sec. 2.1) and wave (Sec. 2.2) equations; then, we discuss their discretization (Sec. 2.3), the selection of the time scale, and the computation of their solution (Sec. 2.4). Finally (Sec. 2.5), we compare different methods for the computation of the solution to the heat equation.

2.1 HEAT EQUATION

The *scale-based representation* $H : \mathcal{N} \times \mathbb{R}^+ \to \mathbb{R}$ of the function $h : \mathcal{N} \to \mathbb{R}$ is the solution to the *heat diffusion equation* $(\partial_t + \Delta)H(\mathbf{p}, t) = 0$, $H(\cdot, 0) = h$. The function $H(\mathbf{p}, t)$ represents the heat distribution at the point $\mathbf{p}$ and at time t, where h is the initial distribution. The solution to the heat equation is written as

$$H(\mathbf{p}, t) = \langle K_t(\mathbf{p}, \cdot), h \rangle_2 = \sum_{n=0}^{+\infty} \exp(-\lambda_n t) \langle h, \phi_n \rangle_2 \phi_n(\mathbf{p}), \qquad (2.1)$$

where $K_t(\mathbf{p}, \mathbf{q}) = \sum_{n=0}^{+\infty} \exp(-\lambda_n t) \phi_n(\mathbf{p}) \phi_n(\mathbf{q})$ is the spectral representation of the *heat diffusion kernel*. The heat diffusion and the Laplace-Beltrami operators have the same eigenfunctions $\{\phi_n\}_{n=0}^{+\infty}$ and $(\exp(-\lambda_n t))_{n=0}^{+\infty}$ are the eigenvalues of the heat operator. The heat kernel is invariant to isometries and verifies the *semi-group* $\langle K_{t_1}, K_{t_2} \rangle_2 = K_{t_1+t_2}$ and *inversion* $K_t^{-1} = K_{-t}$ properties.

The spectral representation (2.1) shows the smoothing effect on the initial condition h; as the scale increases, the component of h along the eigenfunctions associated with the larger Laplacian eigenvalue becomes null. We also notice that the normalized function $\mathcal{A}_{\mathcal{N}}^{-1} H(\cdot, t)$ with respect to the surface area $\mathcal{A}_{\mathcal{N}}$ minimizes the weighted least-squares error $\int_{\mathcal{N}} K_t(\mathbf{p}, \mathbf{q}) |h(\mathbf{q}) - g(\mathbf{p})|^2 d\mathbf{q}$ on $\mathcal{L}_2(\mathcal{N})$, for a given h. A generalization of these properties for the Laplacian spectral kernel is discussed in Sec. 3.2.

We now introduce the main properties of the heat kernel on surfaces and volumes (Sec. 2.1.1) and the selection of the optimal time scale of the heat kernel (Sec. 2.1.2); then, we discuss the comparison of the heat kernel at different scales (Sec. 2.1.3).

2.1.1 HEAT EQUATION ON SURFACES AND VOLUMES

On surfaces, the heat kernel satisfies the following properties [SOG09, Gri06]:

- for an isometry $\Phi : \mathcal{N} \to \mathcal{Q}$ between two manifolds $\mathcal{N}, \mathcal{Q}$,

$$K_t^{\mathcal{N}}(\mathbf{p}, \mathbf{q}) = \mathcal{K}_t^{\mathcal{Q}}(\Phi(\mathbf{p}), \Phi(\mathbf{q})), \quad \forall \mathbf{p}, \mathbf{q} \in \mathcal{N}, \forall t \in \mathbb{R}^+; \tag{2.2}$$

- if Φ is surjective and Eq. (2.2) holds, then Φ is an isometry;

- if D is a compact set of $\mathcal{N}$, then

$$\lim_{t \to 0} K_t^D(\mathbf{p}, \mathbf{q}) = K_t^{\mathcal{N}}(\mathbf{p}, \mathbf{q});$$

- if $D_1 \subseteq D_2 \subseteq \mathcal{N}$, then $K_t^{D_1}(\mathbf{p}, \mathbf{q}) \leq K_t^{D_2}(\mathbf{p}, \mathbf{q})$; and

- on smooth and polygonal surfaces, the heat kernel fully determines the Riemannian metric [ZGLG12].

For small values of t [SOG09, Var67], the auto-diffusivity function

$$K_t(\mathbf{p}, \mathbf{p}) \approx \begin{cases} (4\pi t)^{-1}(1 + 1/3t\kappa(\mathbf{p})) + \mathcal{O}(t^2), \\ (4\pi t)^{3/2}(1 + 1/6s(\mathbf{p})), \end{cases} \quad t \to 0$$

encodes the Gaussian $\kappa(\mathbf{p})$ and total $s(\mathbf{p})$ curvature at $\mathbf{p}$. For large t, $K_t(\mathbf{p}, \mathbf{p})$ is dominated by the Fiedler vector ϕ_1 [Fie73] (i.e., the first non-trivial eigenfunction), which encodes the global structure of the input shape. According to [SOG09, dGGV08], the surface $\mathcal{N}$ at $\mathbf{p}$ can be characterized in terms of the average squared diffusion distance at $\mathbf{p}$ (*eccentricity*), which is defined as

$$\text{ecc}_t(\mathbf{p}) = \mathcal{A}_{\mathcal{N}}^{-1} \int_{\mathcal{N}} d_t(\mathbf{p}, \mathbf{q}) d\mathbf{q} = K_t(\mathbf{p}, \mathbf{p}) + E_{\mathcal{N}}(t) - 2\mathcal{A}_{\mathcal{N}}^{-1}, \quad t \to 0,$$

where $E_{\mathcal{N}}(t) := \sum_{n=0}^{+\infty} \exp(-\lambda_n t)$ is the sum of the eigenvalues of the heat kernel. Since the area and trace are independent of the evaluation point, the functions ecc_t and $K_t(\mathbf{p}, \cdot)$ have the same level sets and extrema on $\mathcal{N}$. In particular, for small scales the extrema of the eccentricity are localized at the curvature extrema.

On volumes, the analytical representation of the (volumetric) heat kernel $K_t(\mathbf{p}, \mathbf{q}) := (4\pi t)^{-3/2} \exp(-\|\mathbf{p} - \mathbf{q}\|_2^2/4t)$ allows us to solve the heat equation as $F(\cdot, t) = \langle K_t, f \rangle_2$ and without computing the Laplacian spectrum (Sec. 2.4.4). We briefly recall that the volumetric heat kernel has been applied also to the discretization of the Laplace-Beltrami operator on point sets (Sec. 1.1.2).

2.1.2   OPTIMAL TIME VALUE OF THE HEAT KERNEL

As optimal time value, we select the scale that provides a small residual error $\|F(\cdot, t) - f\|_2$ and a low energy $\|F(\cdot, t)\|_2$, which controls the solution smoothness. Through the Laplacian spectrum

$(\lambda_n, \phi_n)_{n=0}^{+\infty}$, the orthonormality of the Laplacian eigenfunctions, and the spectral representation $f = \sum_{n=0}^{+\infty} \langle f, \phi_n \rangle_2 \phi_n$ of the initial condition, we rewrite these terms as

$$
\begin{cases}
\|F(\cdot, t) - f\|_2^2 = \sum_{n=0}^{+\infty} |1 - \exp(-2\lambda_n t)|^2 |\langle f, \phi_n \rangle_2|^2, \\
\|F(\cdot, t)\|_2^2 = \sum_{n=0}^{+\infty} \exp(-2\lambda_n t) |\langle f, \phi_n \rangle_2|^2;
\end{cases}
$$

indeed, the residual and penalty terms are increasing and decreasing maps with respect to t, respectively. If t tends to zero, then the residual becomes null and the energy converges to $\|f\|_2$. If t becomes large, then the residual tends to $|\langle f, \phi_0 \rangle_2|$ and the solution norm converges to $(\|f\|_2^2 - |\langle f, \phi_0 \rangle_2|^2)^{1/2}$. According to these properties, the plot (L-curve) of $\epsilon(t) := (\|F(\cdot, t) - f\|_2, \|F(\cdot, t)\|_2)$ is L-shaped [HO93] and its minimum provides the optimal time value, i.e., the best compromise between approximation accuracy and smoothness. For the computation of the optimal time value (Fig. 2.1), we apply the corner detection based on cubic B-splines approximation [HO93]; alternatives are the evaluation of the curvature of the graph of $\epsilon(t)$ or its adaptive pruning [HO93]. Finally, we mention the analogy between the selection of the optimal scale value for the heat kernel and the selection of the regularization parameter for the Laplacian smoothing of noisy scalar functions (Sec. 5.2.2).

2.1.3 COMPARISON OF THE HEAT KERNEL AT DIFFERENT SCALES

Let us study the variation of the $\mathcal{L}_2$ norm of the heat kernels centerd at the same point and associated with two different scales. This variation is expressed in terms of the Laplacian spectrum as (Fig. 2.2)

$$
\|K_s(\mathbf{p}, \cdot) - K_t(\mathbf{p}, \cdot)\|_2^2 = \sum_{n=0}^{+\infty} |\exp(-\lambda_n s) - \exp(-\lambda_n t)|^2 |\phi_n(\mathbf{p})|^2, \tag{2.3}
$$

and its behavior is mainly guided by the function

$$
F_{s,t}(\lambda) := |\exp(-\lambda s) - \exp(-\lambda t)|^2;
$$

in particular, $F_{s,t}(0) = 0$ and $\lim_{\lambda \to +\infty} F_{s,t}(\lambda) = 0$. The derivative of $F_{s,t}$ with respect to λ is

$$
\partial_\lambda F_{s,t}(\lambda) = 2 \left[\exp(-\lambda s) - \exp(-\lambda t)\right] \left[-s \exp(-\lambda s) + t \exp(-\lambda t)\right]
$$

and it vanishes if and only if $s = t$ or $\lambda = \lambda_0 := \frac{\log t - \log s}{t - s}$. Indeed, the unique maximum of $F_{s,t}$ is attained at λ_0 and its value is $F_0 := F_{s,t}(\lambda_0) = \exp(-1) - \exp(-t/s)$. Assuming that $s < t$, the behavior of λ_0 and $F_{s,t}(\lambda_0)$ with respect to s, t and their ratio is guided by the following relations:

$$
\lim_{t \to 0} \lambda_0 = -\infty, \qquad \lim_{t \to s} \lambda_0 = s^{-1}, \quad \lim_{t \to +\infty} \lambda_0 = 0^+,
$$
$$
\lim_{t/s \to 0} F_0 = (e^{-1} - 1)^2, \quad \lim_{t/s \to 1} F_0 = 0, \quad \lim_{t/s \to +\infty} F_0 = e^{-2}.
$$

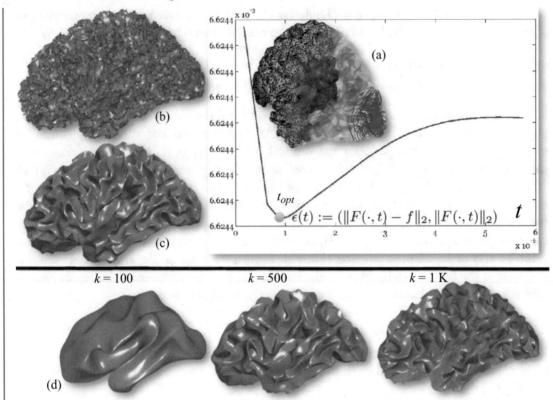

Figure 2.1: (a) Input mesh and L-curve of the approximation accuracy (y-axis) vs. the solution smoothness (x-axis). (b) Data set achieved by adding a Gaussian noise to (a). Diffusion smoothing computed with (c) the Padé-Chebyshev approximation ($r = 7$), and (d) the truncated approximation with k Laplacian eigenparis. A lower number of eigenpairs smooths local details; increasing k reconstructs the noisy component. The ℓ_∞ error between the ground-truth (a) and the smooth approximation of (b) is lower than 1% for the Padé-Chebyshev method (c) and varies from 12% ($k = 100$) to 13% ($k = 1K$) for the truncated approximation (d).

2.2 WAVE EQUATION

The heat equation is strictly related to the *Schroedinger (wave) equation* $(\partial_t + i\Delta)H(\cdot, t) = 0$, with initial condition $H(\cdot, 0) = h$, which represents the physical model of a quantum particle with initial energy h. The spectral representation of the solution is $H(\cdot, t) = \sum_{n=0}^{+\infty} \exp(-i\lambda_n t)\langle h, \phi_n\rangle_2 \phi_n$, i.e., a complex wave function with oscillatory behavior. This periodic effect is due to the real and complex parts of the filter $\exp(i\lambda_n t) = \cos(\lambda_n t) + i\sin(\lambda_n t)$. The norm of the solution is the probability $P_t(\mathbf{p})$ to

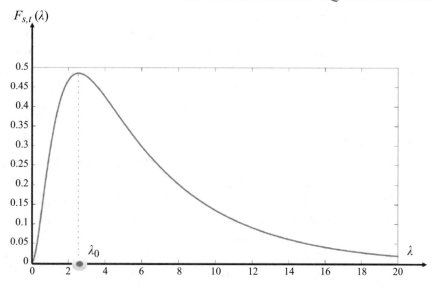

Figure 2.2: Behavior of the variation of the $\mathcal{L}_2$ norm of the heat kernel (c.f., Eq. (2.3)).

find a point $\mathbf{p}$ after a time t; in fact, the following identity holds

$$P_t(\mathbf{p}) = \lim_{T \to +\infty} \int_0^T |H(\mathbf{p}, t)|^2 dt$$
$$= \sum_{n=0}^{+\infty} |\langle h, \phi_n \rangle_2|^2 |\phi_n(\mathbf{p})|^2$$
$$= \|H(\mathbf{p}, t)\|_2^2.$$

Finally, the heat equation is related to the *mean curvature flow* [CPS13, KSB12] $(\partial_t + \Delta_t)\Phi_t = 0$, where $\Phi_t : \mathcal{M} \to \mathbb{R}^3$ is a family of immersions and Δ_t is the Laplace-Beltrami operator associated with the metric induced by the immersion at time t. For details on the link between the Laplace-Beltrami operator and the Ricci flow, we refer the reader to [ZZG+15].

2.3 DISCRETE HEAT EQUATION AND KERNEL

We briefly introduce the weak formulation [All07] of the heat equation; similar results apply to the equations previously introduced. Chosen a set $\mathcal{B} := \{\psi_i\}_{i=1}^n$ of linearly independent functions on $\mathcal{N}$, we approximate the solution $\tilde{F}(\mathbf{p}, t) := \sum_{i=1}^n a_i(t)\psi_i(\mathbf{p})$ to the weak heat equation as $\langle \partial_t \tilde{F}(\cdot, t), \psi_i \rangle_2 + \langle \Delta \tilde{F}(\cdot, t), \psi_i \rangle_2 = 0$, $i = 1, \ldots, n$. Introducing the matrices $\mathbf{L} := (\langle \Delta \psi_i, \psi_j \rangle_2)_{i,j=1}^n$ and $\mathbf{B} := (\langle \psi_i, \psi_j \rangle_2)_{i,j=1}^n$, the discrete heat equation becomes $(\mathbf{B}\partial_t + \mathbf{L})\mathbf{a}(t) = \mathbf{0}$, $\mathbf{a}(t) := (a_i(t))_{i=1}^n$. An analogous relation can be derived for the boundary condition $F(\mathbf{p}, 0) = h(\mathbf{p})$. Since $\mathbf{B}$ is the Gram matrix associated with $\mathcal{B}$, it is invertible and the previous system of equations is $(\partial_t + \mathbf{B}^{-1}\mathbf{L})\mathbf{a}(t) = \mathbf{0}$, with initial condition $\mathbf{F}(0) = \mathbf{f}$. Then, the solution to the discrete heat equation is expressed as a linear combination of the Laplacian eigensystem as $\mathbf{F}(t) = \sum_{i=1}^{n} \exp(-\lambda_i t)\langle \mathbf{f}, \mathbf{x}_i \rangle_{\mathbf{B}} \mathbf{x}_i$. We now discuss the main properties of the discrete heat kernel (Sec. 2.3.1), the linear independence of the heat kernel at different scales and points (Sec. 2.3.2).

2.3.1 PROPERTIES OF THE HEAT KERNEL

The solution to the discrete heat equation is $\mathbf{F}(t) = \mathbf{K}_t \mathbf{f}$ (Fig. 2.3), where $\mathbf{K}_t := \mathbf{X}\mathbf{D}_t\mathbf{X}^\top\mathbf{B}$, $\mathbf{D}_t := \text{diag}\,(\exp(-\lambda_i t))_{i=1}^n$, is the *heat kernel matrix* (Table 2.1). Lumping the linear FEM mass matrix $\mathbf{B}$ to the diagonal matrix $\mathbf{D}$, whose entries are the areas of the Voronoi regions, the heat kernel becomes equal to the *Voronoi-cotangent heat kernel* $\mathbf{K}_t^\star := \mathbf{X}\mathbf{D}_t\mathbf{X}^\top\mathbf{D}$, $\mathbf{L}\mathbf{X} = \mathbf{X}\Lambda$. Choosing $\mathbf{B} := \mathbf{I}$, we get the *heat kernel* $\tilde{\mathbf{K}}_t := \mathbf{X}\mathbf{D}_t\mathbf{X}^\top$ with cotangent weights. Analogously to the results in Sec. 2.1, the heat kernel matrix satisfies the following relations: $\mathbf{K}_{t_1}\mathbf{K}_{t_2} = \mathbf{K}_{t_2}\mathbf{K}_{t_1} = \mathbf{K}_{t_1+t_2}$ (*commutative* and *semi-group properties*), $\mathbf{K}_t^{-1} = \mathbf{K}_{-t}$ (*inversion property*). Finally, we recall that the discrete heat kernel determines the discrete Reimannian metric [ZGLG12].

Figure 2.3: Anisotropic heat kernel centerd at a (black) seed point on a coarse triangle mesh.

Table 2.1: Main properties of the discrete heat kernel: sparsity, positive definiteness, and symmetry. The full • and empty ∘ circle means that the corresponding property is or is not satisfied, respectively.

Heat Ker.	Matrix $\mathbf{K}_t$	Sparsity	Pos. Def.	Sym.	Cov.	Inv.
Std	$\mathbf{X}\mathbf{D}_t\mathbf{X}^\top$	∘	•	•	∘	∘
Vor.-cot	$\mathbf{X}\mathbf{D}_t\mathbf{X}^\top\mathbf{D}$	∘	•	∘	•	∘
wFEM	$\mathbf{X}\mathbf{D}_t\mathbf{X}^\top\mathbf{B}$	∘	•	∘	•	∘

If **B** is the linear FEM mass matrix or the diagonal matrix of the Voronoi areas, then the heat kernel matrix $\mathbf{K}_t$ is intrinsically *scale-covariant*, i.e., rescaling the points of $\mathcal{M}$ by a factor α, $\alpha > 0$, and indicating the new surface as $\alpha\mathcal{M}$, we get that only the time component of the kernel is rescaled. In fact, the rescaling changes the matrix **B** and the eigensystem $\{(\lambda_i, \mathbf{x}_i)\}_{i=1}^n$ of $\mathcal{M}$ into $\alpha^2\mathbf{B}$ and $\{(\alpha^{-2}\lambda_i, \alpha^{-1}\mathbf{x}_i)\}_{i=1}^n$, respectively. Indeed, $\mathbf{K}_t(\alpha\mathcal{M}) = \mathbf{K}_{\alpha^{-2}t}(\mathcal{M})$ without an *a-posteriori* normalization. The scale-covariance of $\mathbf{K}_t$ is guaranteed by the mass matrix, which changes according to the surface rescaling and compensates the variation of the corresponding Laplacian spectrum.

The kernel becomes *scale-invariant* (i.e., $\mathbf{K}_t(\alpha\mathcal{M}) = \mathbf{K}_t(\mathcal{M})$) by normalizing each eigenvalue by λ_n, which is efficiently computed using the inverse method [GV89, VL08]. Alternatively, the scale-invariance and covariance of the heat kernel is achieved in the Fourier domain [BK10]. In [BBB+10, BBC+10], the matching performances of heat kernel descriptors have been tested against shape transformation, sampling, and noise.

2.3.2   LINEAR INDEPENDENCE OF THE HEAT KERNEL AT DIFFERENT POINTS AND SCALES

First of all, we notice that $\mathcal{B} := \{K_t(\mathbf{p}_i, \cdot)\}_{i=1}^n$ are linearly independent functions in $\mathcal{F}(\mathcal{M})$; in fact,

$$0 = \sum_{i=1}^n \alpha_i K_t(\mathbf{p}_i, \cdot) = \mathbf{K}_t\alpha \iff \alpha = 0.$$

Indeed, any function $f \in \mathcal{F}(\mathcal{M})$ can be expressed in terms of $\mathcal{B}$ as $\mathbf{f} = \sum_{i=1}^n \alpha_i K_t(\mathbf{p}_i, \cdot)$ and the coefficients are $\alpha = \mathbf{K}_{-t}\mathbf{f}$.

Let us show that the functions $\mathcal{B} := \{K_{t_i}(\mathbf{p}_j, \cdot)\}_{i=1}^r$ induced by the heat kernel, associated with different scales, and centerd at the same point are linearly independent. In fact,

$$0 = \sum_{i=1}^r \alpha_i \mathbf{X}\mathbf{D}_{t_i}\mathbf{X}^\top\mathbf{B}\mathbf{e}_j$$

$$= \sum_{k=1}^n \left(\sum_{i=1}^r \alpha_i \exp(-\lambda_k t_i)\right) \langle\mathbf{x}_k, \mathbf{e}_j\rangle_\mathbf{B}\mathbf{x}_k$$

if and only if

$$0 = \sum_{i=1}^r \alpha_i \exp(-\lambda_k t_i)\langle\mathbf{x}_k, \mathbf{e}_j\rangle_\mathbf{B}$$

$$= \langle\mathbf{x}_k, \sum_{i=1}^r \alpha_i \exp(-\lambda_k t_i)\mathbf{e}_j\rangle_\mathbf{B}, \quad k = 1, \ldots, n.$$

From the last relation, we get that

$$0 = \sum_{i=1}^{r} \alpha_i \exp(-\lambda_k t_i) e_j, \iff \alpha_j = 0, \quad j = 1, \ldots n.$$

We can now use the basis functions induced by the heat kernel at different points and scales to approximate any new function $\mathbf{K}_t \mathbf{f}$ as $\sum_{i=1}^{r} \alpha_i \mathbf{K}_{t_i} \mathbf{f}$. To this end, we rewrite the least-squares error as

$$\left\| (\mathbf{K}_t - \sum_{i=1}^{r} \alpha_i \mathbf{K}_{t_i}) \mathbf{f} \right\|_{\mathbf{B}}^2 = \mathbf{f}^\top \left[\mathbf{K}_t^\top \mathbf{B} \mathbf{K}_t - 2 \sum_{i=1}^{r} \alpha_i \mathbf{K}_t^\top \mathbf{B} \mathbf{K}_{t_i} + \sum_{i,j=1}^{r} \alpha_i \alpha_j \mathbf{K}_{t_i}^\top \mathbf{B} \mathbf{K}_{t_j} \right] \mathbf{f}$$

whose derivative with respect to α vanishes if

$$0 = 2\mathbf{f}^\top \left[\sum_{i=1}^{r} \alpha_i \mathbf{K}_{t_i}^\top \mathbf{B} \mathbf{K}_{t_l} - \mathbf{K}_t^\top \mathbf{B} \mathbf{K}_{t_l} \right] \mathbf{f}$$

$$= 2\mathbf{f}^\top \left[\sum_{i=1}^{r} \alpha_i \mathbf{B} \mathbf{K}_{t_i + t_l} - \mathbf{B} \mathbf{K}_{t + t_l} \right] \mathbf{f}.$$

In the last equality, we have applied the identity $\mathbf{K}_t^\top \mathbf{B} \mathbf{K}_s = \mathbf{B} \mathbf{K}_{t+s}$. Indeed, the coefficients $\alpha := (\alpha_i)_{i=1}^{r}$ are achieved by solving the $r \times r$ linear system

$$\sum_{i=1}^{r} \alpha_i \mathbf{f}^\top \mathbf{B} \mathbf{K}_{t_i + t_l} \mathbf{f} = \mathbf{f}^\top \mathbf{K}_{t + t_l} \mathbf{f}, \quad l = 1, \ldots, r,$$

whose coefficient matrix and right-hand side vector are computed through the Padé-Chebyshev method.

2.4   COMPUTATION OF THE DISCRETE HEAT KERNEL

For the computation of the solution to the discrete heat equation and kernel, we consider linear (Sec. 2.4.1), polynomial (Sec. 2.4.2), and rational (Sec. 2.4.3) approximations of the exponential filter. On volumes (Sec. 2.4.4), we discuss the solution to the heat equation based on the analytic representation of the heat kernel. With the exception of the truncated spectral method, all the previous approximations are independent of the evaluation of the Laplacian spectrum and reduce to a set of sparse linear systems (Table 2.2). The polynomial and rational approximations generally provide the best compromise between approximation accuracy and computational cost.

2.4.1   LINEAR APPROXIMATION

For the solution to the heat equation, we review the truncated spectral approximation, the Euler backward method, the first-order Taylor approximation, and the Krylov and Schur methods.

Table 2.2: Numerical computation of the solution to the heat equation; $\tau(n)$ is the cost for the solution of a sparse linear system

Method	Numerical Scheme	Scales	Comput. Cost
Linear Approximation			
Truncated spectral approx. [GV89, VBCG10]	$\mathbf{F}_k(t) = \sum_{i=1}^{k} \exp(-\lambda_i t)\langle \mathbf{f}, \mathbf{x}_i\rangle_{\mathrm{B}}\mathbf{x}_i$	Any	$\mathcal{O}(kn)$
Euler backward approx. [CDR00, DMSB99, ZH08]	$(t\tilde{\mathbf{L}} + \mathbf{I})\mathbf{F}_{k+1}(t) = \mathbf{F}_k(t)$	Small	$\mathcal{O}(\tau(n))$
I order Taylor approx. [CDR00, DMSB99]	$\mathbf{B}\mathbf{F}(t) = (\mathbf{B} - t\mathbf{L})f$	Small	$\mathcal{O}(\tau(n))$
Krylov/Schur approx. [GV89, Saa92, ZH08]	Projection on $\{\mathbf{g}_i := (\mathbf{B}^{-1}\mathbf{L})^i \mathbf{f}\}_{i=1}^{m}$	Any	$\mathcal{O}(m\tau(n))$, $\mathbf{B} \neq \mathbf{I}$ $\mathcal{O}(n)$, $\mathbf{B} = \mathbf{I}$
Polynomial Approximation			
Power approx. [GV89]	$\mathbf{F}(t) = \sum_{i=0}^{m} \mathbf{g}_i / i!$ $\mathbf{g}_i := \tilde{\mathbf{L}}^i \mathbf{f}$	Any	$\mathcal{O}(m\tau(n))$, $\mathbf{B} \neq \mathbf{I}$ $\mathcal{O}(n)$, $\mathbf{B} = \mathbf{I}$
Rational Approximation			
Padé-Cheb. approx. [CRV84, Sid98, Saa92, Pat16a]	$\mathbf{F}(t) = \alpha_0 \mathbf{f} + \sum_{i=1}^{r} \mathbf{g}_i$ $(t\,\mathbf{L} + \theta_i\,\mathbf{B})\,\mathbf{g}_i = -\alpha_i\mathbf{B}\,\mathbf{f}$	Any	$\mathcal{O}(r\tau(n))$
Contour integral approx. [Pus11]	$\mathbf{F}(t) = \sum_{i=1}^{r} \alpha_i\,\mathbf{g}_i$ $(\alpha_i)_{i=1}^{r}$ quadr. coeff.	Any	$\mathcal{O}(r\tau(n))$

Truncated spectral approximation and power method   The computational bottleneck for the evaluation of the whole Laplacian spectrum imposes on us to consider only a small subset of the Laplacian spectrum. Since the decay of the *filter factor* $\exp(-\lambda_i t)$ increases with λ_i, in the spectral representation of the solution to the heat equation we consider only the contribution related to the first k eigenpairs, i.e., $\mathbf{F}_k(t) = \sum_{i=1}^{k} \exp(-\lambda_i t)\langle \mathbf{f}, \mathbf{x}_i\rangle_{\mathrm{B}}\mathbf{x}_i$. The truncated approximation is accurate only if the exponential filter decays fast (e.g., large values of time) and the effect of the selected eigenpairs on the approximation accuracy cannot be estimated without computing the whole spectrum. The *multi-resolution prolongation operators* [VBCG10] prolongate the values of the truncated spectral approximation, computed on a low-resolution representation of the input shape, to the initial resolution through a hierarchy of simplified meshes. In this case, the number of eigenpairs are heuristically adapted to the surface resolution and its global/local features.

Euler backward method   In [CDR00, DMSB99], the solution to the heat equation is computed through the Euler backward method $(t\tilde{\mathbf{L}} + \mathbf{I})\mathbf{F}_{k+1}(t) = \mathbf{F}_k(t)$, $\mathbf{F}_0 = \mathbf{f}$. The resulting functions are over-smoothed and converge to a constant function, as $k \to +\infty$.

First-order Taylor approximation and power method   Since the derivative of $\mathbf{K}_t$ at $t = 0$ equals the Laplacian matrix (i.e., $(\mathbf{I} - \mathbf{K}_t)/t \to \mathbf{B}^{-1}\mathbf{L}$, $t \to 0$), the heat kernel $\mathbf{K}_t$ is approximated by $(\mathbf{I} - t\mathbf{B}^{-1}\mathbf{L})$ and $\mathbf{F}(t) = \mathbf{K}_t\mathbf{f}$ solves the sparse linear system $\mathbf{B}(\mathbf{K}_t\mathbf{f}) = (\mathbf{B} - t\mathbf{L})\mathbf{f}$. This last relation gives an approximation of $\mathbf{F}(t)$ that is independent of the Laplacian spectrum and is valid only for small values of t. For an arbitrary value of t, the "power" method applies the identity $(\mathbf{K}_{t/m})^m = \mathbf{K}_t$, where m is chosen in such a way that t/m is sufficiently small to guarantee that the approximation $\mathbf{K}_{t/m} \approx (\mathbf{I} - t/m\tilde{\mathbf{L}})$ is accurate. However, the selection of m and its effect on the approximation accuracy cannot be estimated a-priori.

Krylov and Schur approximations   The Krylov subspace projection [GV89, Saa92] computes an approximation of $\exp(-t\mathbf{A})\mathbf{f}$ in the space generated by the vectors $\mathbf{f}, \mathbf{A}\mathbf{f}, \ldots, \mathbf{A}^{m-1}\mathbf{f}$, thus processing a $m \times m$ matrix instead of a $n \times n$ matrix, where m is much lower than n (e.g., $m \approx 20$). This approximation [ZH08] becomes computationally expensive when the dimension of the Krylov space increases, still remaining much lower than n (e.g., $n \approx 5K$). In both cases, the vector $\tilde{\mathbf{L}}^i\mathbf{f} = (\mathbf{B}^{-1}\mathbf{L})^i\mathbf{f}$ must be computed without inverting the mass matrix; to this end, we notice that the vector $\mathbf{g}_i := (\mathbf{B}^{-1}\mathbf{L})^i\mathbf{f}$ satisfies the linear system $\mathbf{B}\mathbf{g}_i = \mathbf{L}\mathbf{g}_{i-1}$, $\mathbf{B}\mathbf{g}_1 = \mathbf{L}\mathbf{f}$. Since the coefficient matrix $\mathbf{B}$ is sparse, symmetric, and positive definite, the vectors $(\mathbf{g}_i)_{i=1}^m$ are evaluated in linear time by applying iterative solvers (e.g., conjugate gradient) or pre-factorizing $\mathbf{B}$.

2.4.2   POLYNOMIAL APPROXIMATION

The exponential of a matrix $\mathbf{A}$ is defined as the exponential *power series* $\exp(\mathbf{A}) = \sum_{n=0}^{+\infty} \mathbf{A}^n/n!$, which converges for any square matrix $\mathbf{A}$. Even though the input matrix $\mathbf{A}$ is sparse, its exponential $\exp(-t\mathbf{A})$ is always full ($t \neq 0$) and can be computed or stored only if $\mathbf{A}$ has a few hundred rows and columns only. In particular, for computer graphics applications we can consider 3D shapes only with a small number of samples (i.e., few hundreds) or evaluate the heat kernel on a set of seed points that are representative of the geometry and features of the input shape.

2.4.3   RATIONAL APPROXIMATION

The exponential of an arbitrary matrix $\mathbf{A}$ is equal to the complex contour integral $\exp(t\mathbf{A}) = (2\pi i)^{-1} \int_\gamma \exp(z)(z\mathbf{I} - t\mathbf{A})^{-1}dz$, where γ is a closed contour winding once around the spectrum of $t\mathbf{A}$ [GV89, Sec. 11], [Rud87, Sec. 10]. From this identity, we introduce two accurate and computationally efficient approximations of the exponential of the Laplacian matrix.

Padé-Chebyshev approximation   The rational approximation of the exponential function of order (k, k) and with simple poles is $r_{kk}(z) := p_k(z)/q_k(z) = \alpha_0 + \sum_{i=1}^k \alpha_i(z - \theta_i)^{-1}$, where p_k, q_k are polynomials of order k, $\alpha_0 = \lim_{z \to +\infty} r_{kk}(z)$, θ_i is a pole, and α_i is the residual at θ_i. Applying this last relation to $t\mathbf{A}$, we get $\exp(t\mathbf{A}) = \alpha_0\mathbf{I} + \sum_{i=1}^k \alpha_i(t\mathbf{A} - \theta_i\mathbf{I})^{-1}$. Among the rational approximations of the exponential function, we focus on its best approximation $r_{kk}(\cdot)$

of order k with respect to the ℓ_∞ norm, i.e., the unique $r_{kk}(z) = p_k(z)/q_k(z)$ that minimizes the error $\|\pi(z) - \exp(-z)\|_\infty$ in the space $\Pi_{kk} \ni \pi$ of rational polynomials of order k.

Here, the main difficulty is the evaluation of the coefficients and poles of the rational approximation of the exponential function for a given k, which is generally affected by the ill-conditioned computation of the polynomial roots. These coefficients and poles have been computed with a different accuracy and for different orders of the rational polynomial [CRV84, CMV69, MVL03, Sid98, Saa92]. These approximations are also included in standard numerical libraries for signal processing. Finally, we recall that in spectral graph theory [OSV12], the Padé-Chebyshev and the Lanczos methods have been applied to the approximation of $\exp(-\mathbf{A})\mathbf{f}$, where $\mathbf{A}$ is a symmetric and positive semi-definite matrix.

The idea behind the *spectrum-free computation* [Pat13, Pat14] is to apply the (r, r)-degree Padé-Chebyshev rational approximation to the exponential representation $\mathbf{F}(t) = \exp(-t\tilde{\mathbf{L}})\mathbf{f}$ of the solution to the heat equation $(\partial_t + \tilde{\mathbf{L}})\mathbf{F}(t) = \mathbf{0}$, $\mathbf{F}(0) = \mathbf{f}$ (Algorithm 2.1). In this case, the solution $\mathbf{F}(t) = \alpha_0 \mathbf{f} + \sum_{i=1}^r \mathbf{g}_i$ is the sum of the solutions of r sparse linear systems $(t\mathbf{L} + \theta_i \mathbf{B})\mathbf{g}_i = -\alpha_i \mathbf{B}\mathbf{f}$, $i = 1, \ldots, r$. The resulting approximation belongs to the linear space generated by $\mathbf{f}$ and $\{\mathbf{g}_i\}_{i=1}^r$, which are calculated as a minimum norm residual solution [GV89], depend on the input domain, the initial condition, and the selected time value. In comparison, the Laplacian eigenfunctions only encode the domain geometry and it is difficult to select the number of eigenpairs necessary to achieve a given approximation of $\mathbf{F}(t)$ with respect to t and $\mathbf{f}$.

Algorithm 2.1 Padé-Chebyshev approximation of the solution to the heat equation.

Require: A function $f : \mathcal{P} \to \mathbb{R}$, $\mathbf{f} := (f(\mathbf{p}_i))_{i=1}^n$.
Ensure: The approximate solution $\mathbf{F}(t) = \mathbf{K}_t \mathbf{f}$ of $\mathbf{f}$ to the heat equation.
 1: Select the value of t (e.g., optimal value, Sec. 2.1.2).
 2: **for** $i = 1, \ldots, r - 1$ **do**
 3:    Compute $\mathbf{g}_i$: $(t\mathbf{L} + \theta_i \mathbf{B})\mathbf{g}_i = -\alpha_i \mathbf{B}\mathbf{f}$.
 4: **end for**
 5: Approximate $\mathbf{K}_t \mathbf{f}$ as $\alpha_0 \mathbf{f} + \sum_{i=1}^r \mathbf{g}_i$.

This approximation is independent of the computation of the Laplacian spectrum, user-defined parameters, and multi-resolutive prolongation operators [VBCG10], which heuristically adapt the number of eigenpairs to the surface resolution. The sparse and well-conditioned matrices of the previous linear systems have the same structure and sparsity of the connectivity matrix of the input domain, properly encode the local and global features in the heat kernel, and can be computed for any representation of the input domain and for any choice of the Laplacian weights. Finally, the accuracy of the Padé-Chebychev approximation is lower than 10^{-r} (e.g., $r = 5, 7$).

The value of t influences the conditioning number of the matrices $(t\mathbf{L} + \theta_i \mathbf{B})$, $i = 1, \ldots, r$. Experiments (Fig. 2.4, [Pat14]) have shown that the linear systems associated with

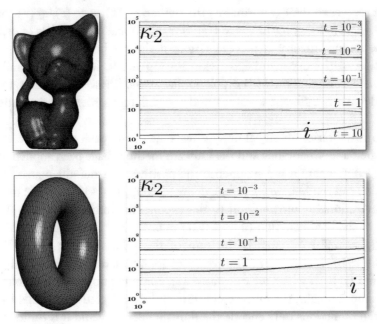

Figure 2.4: Conditioning number κ_2 (y-axis) of the matrices $\{(t\mathbf{L} + \theta_i\mathbf{B})\}_{i=1}^7$, for several values the time parameter t; the indices of the coefficients $\{\theta_i\}_{i=1}^7$ are reported on the x-axis.

the Padé-Chebyshev approximation are generally well conditioned; in any case, pre-conditioners and regularization techniques [GV89] can be applied to attenuate numerical instabilities. Finally, timings on surfaces and volumes (Fig. 2.5) are reduced from 20 up to 1200 times with respect to the approximation based on a fixed number of Laplacian eigenpairs. Laplacian eigenvectors have been computed with the Arnoldi iteration method [LS96, Sor92].

Rational approximation from contour integrals   Since the exponential factor rapidly decays to zero as $Re(z) \to +\infty$, in [Pus11] the complex contour integral has been efficiently computed with quadrature rules. In this case, $\alpha_0 = 0$, the poles $\theta_i := \phi(x_i)$ are evaluated at the quadrature points $\{x_i\}_i, \alpha_i := -(2\pi i)^{-1}h\exp(\phi(x_i))\phi'(x_i)$ are the weights of the quadrature rules, and h is the interval length in the quadrature scheme. The resulting approximation accuracy is guided by the degree of the quadrature rule; low degrees (e.g., $k = 2, k = 4$) generally provide a satisfactory approximation accuracy.

2.4.4   SPECIAL CASE: HEAT EQUATION ON VOLUMES

On a volume, the function $\mathbf{F}(t) = \sum_{i=1}^n \alpha_i V_i K_t(\mathbf{p}_i, \cdot)f(\mathbf{p}_i)$ is approximated as a linear combination of the basis functions $\{K_t(\mathbf{p}_i, \cdot)\}_{i=1}^n$. Here, $\mathbf{V} = \text{diag}(V_i)_{i=1}^n$ is the diagonal matrix

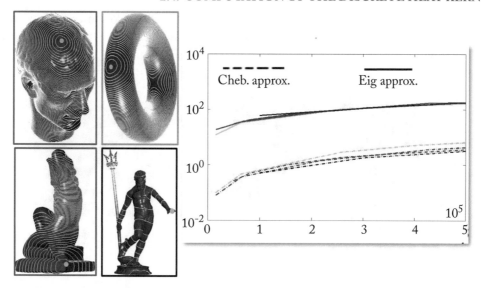

Figure 2.5: Cost (in seconds, y-axis, log-scale) for the evaluation of the diffusion distances on 3D shapes with n samples (x-axis), approximated with $k = 500$ eigenpairs and the Padé-Chebyshev approximation. Colors from the source (orange) point vary from blue (null distance) to red (maximum distance).

of the volumes V_i at $\mathbf{p}_i$, $\mathbf{K}_t$ is the Gram matrix for the Gaussian kernel, and the unknowns $\alpha = (\alpha_i)_{i=1}^{n}$ are determined by imposing the condition $F(\mathbf{p}_i, 0) = f(\mathbf{p}_i), i = 1, \ldots, n$. To overcome the time-consuming solution of the $n \times n$ linear system $\mathbf{VK}_t\alpha = \mathbf{f}$, the number of conditions is reduced or the coefficient matrix is sparsified according to the exponential decay of its entries. Alternatively, the volumetric heat equation is solved by discretizing the Laplace-Beltrami operator with finite elements [All07, RWSN09], or with finite differences on a 6-neighbor stencil [LBB11, LBB12, RBBK10], or with a geometry-driven approximation of the gradient field [LTDZ09, TLHD03].

While a discretization of the heat kernel on a voxel grid is accurate enough for the evaluation of diffusion descriptors [LBB11, RBBK10], which are quantized and clustered in bags-of-features, the computation of the solution to the volumetric heat equation generally requires a more accurate discretization of the input domain. The prolongation of the Laplacian [Rus11a, Rus11b], harmonic [LXW+10, MCK08], and diffusion functions from the volume boundary to its interior, through barycentric coordinates or nonlinear approximation, achieves a low accuracy of the solution in a neighbor of the boundary. The multi-resolution simplification of the input volume is also time-consuming, and the selection of the volume resolution with respect to the expected approximation accuracy are generally guided by heuristics. Indeed, these methods do not intend to approximate the heat kernel quantitatively, but provide alternative approaches that qualitatively behave like the heat kernel on volumes. To improve the accuracy, we consider the volumetric Laplacian matrix of the input domain and compute the Padé-Chebyshev approximation of the induced heat kernel (Fig. 2.6).

$t = 0.1$

$t = 1$

Figure 2.6: Volumetric heat kernel ($r = 7$). Level sets correspond to iso-values uniformly sampled in the range of the solution restricted to the volume boundary.

2.5   DISCUSSION

For the evaluation of the approximation accuracy, we compare the ground-truth distances on cylinders and spheres with the proposed approach and the truncated spectral approximation, where the Laplacian eigenpairs have been computed through a variant [VL08] of the Arnoldi iteration method [LS96, Sor92]. According to [RBG^{+}09], we consider a rectangular domain with edge length $a = 1$, $b = 2$ and the corresponding isometric cylinder ($\cos x, y, \sin x$). Introducing Neumann boundary conditions, their Laplacian eigenpairs are $\phi_{m,n}(x, y) := (\cos(\frac{m\phi}{a}x), \cos(\frac{n\phi}{b}y))$, $m, n \in \mathbb{N}$, and $\lambda_{m,n} := \pi^2(m^2/a^2 + n^2/b^2)$. For the sphere, the spherical harmonics are $\phi_{m,n}(\theta, \varphi) = N \exp(im\varphi) p_l^m(\cos\theta)$, where N is a normalization constant and $p_l^m(\cdot)$ is an associated Legendre function. Since we have an infinite number of eigenpairs, we select k such that the spectral distance $d_k(\mathbf{p}, \mathbf{q}) := \sum_{n=0}^{k} \frac{|\phi_n(\mathbf{p}) - \phi_n(\mathbf{q})|^2}{\varphi^2(\lambda_n)}$ becomes stationary, i.e., $|d_{k+1}(\mathbf{p}, \mathbf{q}) - d_k(\mathbf{p}, \mathbf{q})| < \epsilon$, where ϵ is equal to the 1%.

Figure 2.7 reports the ℓ_∞ error (y-axis) between the ground-truth distances induced by four filters and their approximation with the truncated spectral method with k Laplacian eigenpairs (x-axis) and our approach. For filters with a fast growth (e.g., $\varphi_1 = s^2 \exp(st)$, $\varphi_2 = s \exp(st)$), the truncated spectral approximation provides a good accuracy (i.e., lower than 10^{-5}, with $k \geq 85$ for the cylinder, and $k \geq 137$ for the sphere). Slowly increasing filters generally require a large number of eigenpairs (i.e., $k \geq 300$ for the cylinder, $k \geq 1K$ for the sphere) to achieve an accuracy lower than 10^{-1}.

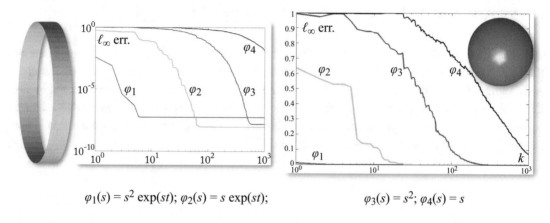

$$\varphi_1(s) = s^2 \exp(st); \quad \varphi_2(s) = s \exp(st); \qquad \varphi_3(s) = s^2; \quad \varphi_4(s) = s$$

Figure 2.7: ℓ_∞ error (y-axis) between the ground-truth distances induced by the filters $\{\varphi_i\}_i$ and the truncated approximation with k (x-axis) eigenpairs. For the Padè-Chebyshev method ($r = 5$) and all the filters, the ℓ_∞ error with respect to the ground-truth is lower than 6.5×10^{-6}.

We also compare the accuracy of the diffusion distances computed with: (i) the Padè-Chebyshev; (ii) the spectral representation of the heat kernel $\mathbf{K}_t$, with k eigenpairs; (iii) the Euler backward method; and (iv) the power method. For all the scales (Fig. 2.8), the accuracy of the Padè-Chebyshev method is higher than the truncated approximation with k eigenpairs, $k = 1, \ldots, 10^3$, the Euler backward method, and the power method. Reducing the scale, the accuracy of the Padè-Chebyshev remains almost unchanged while the other methods are affected by a larger discrepancy and tend to have an analogous behavior ($t = 10^{-4}$). Finally, the Euler backward method generally over-smooths the solution, which converges to a constant as $k \to +\infty$, and the selection of the power m is guided by heuristics.

We discuss the stability of our approach with respect to shape changes or to the approximation of the spectral kernel. To this end, we consider the solution $\mathbf{K}_t \mathbf{e}_i$ to the heat diffusion process, whose initial condition takes value 1 at the anchor point $\mathbf{p}_i$ and 0 otherwise. On noisy

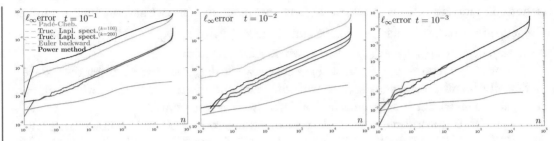

Figure 2.8: ℓ_∞ error (y-axis) between the ground-truth diffusion distances on the cylinder, with a different sampling (x-axis). For different scales, the accuracy of the Padé-Chebyshev method ($r = 5$, orange) remains almost unchanged and higher than the truncated approximation with 100 and 200 eigenpairs (red, blue), the Euler backward (green) and power (black) methods.

and irregularly sampled meshes (Fig. 2.9, Fig. 2.10) or point sets (Fig. 2.11), the level sets of $\mathbf{K}_t\mathbf{e}_i$ are smooth, well-distributed around the anchor point $\mathbf{p}_i$, and remain almost unchanged and coherent with respect to the original shape. These results confirm the robustness of the Padé-Chebyshev approximation to surface discretization. In these examples, the maximum variation between the spectral distances on the complete surface and its representation with holes is lower than 10^{-3}. Additional examples will be discussed in Sec. 4.4.

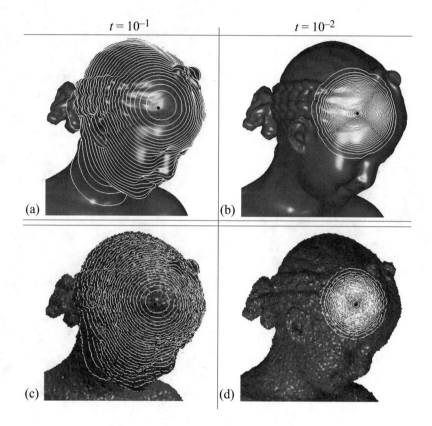

Figure 2.9: Robustness of the Padé-Chebyshev approximation ($r = 7$) of the (a,c) diffusion kernel $\mathbf{K}_t \mathbf{e}_i$ and (b,d) distance at $\mathbf{p}_i$ (black dot) on a smooth and noisy triangulated surface.

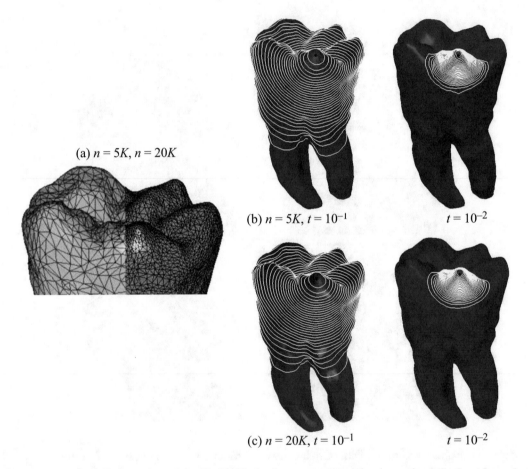

(a) $n = 5K, n = 20K$

(b) $n = 5K, t = 10^{-1}$ $t = 10^{-2}$

(c) $n = 20K, t = 10^{-1}$ $t = 10^{-2}$

Figure 2.10: (b,c) Robustness of the Padé-Chebyshev approximation ($r = 5$) of the heat kernel at different scales ($t = 10^{-1}$, 10^{-2}) with respect to (a) surface sampling ($n = 5K$, $20K$).

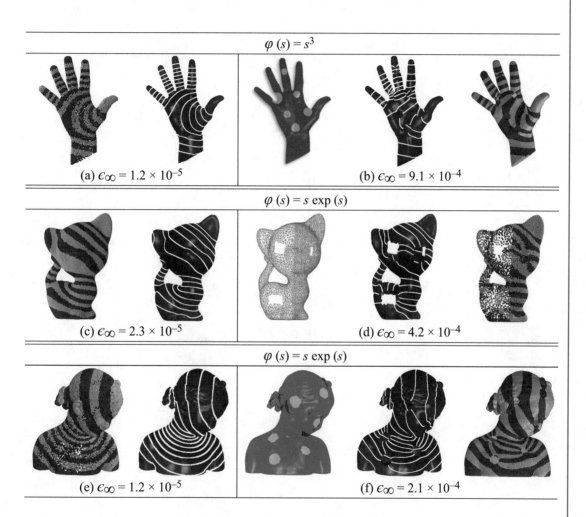

Figure 2.11: Distances computed with the Padé-Chebyshev method ($r = 5$) on (a,c,e) regularly sampled and (b,d,f) irregularly sampled (left) meshes and (right) point sets with holes. To improve the visualization, points are represented as spheres.

CHAPTER 3

Laplacian Spectral Distances

In geometry processing and shape analysis, several applications (e.g., surface remeshing, skeletonization, segmentation, comparison) have been addressed through the definition of shape descriptors and distances. Shape kernels, distances, and descriptors can be defined on 3D shapes by applying:

- *topological and geometric approaches*, which consider the geometric (e.g, curvature) and topological (e.g., genus, connected components) properties of the input shape. As main examples, we mention the Reeb graph, curvature-based descriptors, and geodesic distances; and

- *functional approaches*, which study the differential properties of the space of scalar functions defined on the input shape. As main examples, we mention spherical harmonics, bi-harmonic, diffusion, and wave kernel distances.

We focus on functional approaches for the definition of spectral kernels, distances, and descriptors by filtering the Laplacial spectrum associated with the input shape, i.e., by applying a filter $\varphi : \mathbb{R}^+ \to \mathbb{R}$ to the Laplace-Beltrami operator and to its spectrum. Through the proper selection of the filter, the corresponding spectral kernels, distances, and descriptors satisfy important properties, such as:

- *intrinsic definition*, as a consequence of the use of the Laplacian spectrum, and multi-scale encoding of local and global shape properties, according to the decay of the filter;

- *invariance* to uniform scaling and isometric transformations, which is important for shape analysis and comparison;

- *easy computation* through the evaluation of the Laplacian eigenpairs or the solution of sparse and symmetric linear systems; and

- *approximation of geodesic and optimal transportation distances* on unstructured 3D and n-dimensional data (e.g., diffusion distances in Manifold Learning).

For geometry processing and shape analysis, the distance satisfies the following properties:

- *positivity*: $d(\mathbf{p}, \mathbf{q}) \geq 0$;

- *nullity*: $d(\mathbf{p}, \mathbf{q}) = 0$ if and only if $\mathbf{p} \equiv \mathbf{q}$;

- *symmetry*: $d(\mathbf{p}, \mathbf{q}) = d(\mathbf{q}, \mathbf{p})$; and

- *triangular inequality*: $d(\mathbf{p}, \mathbf{q}) \leq d(\mathbf{p}, \mathbf{r}) + d(\mathbf{r}, \mathbf{q})$.

The spectral distance should be *multi-scale* and *geometry-aware*, through the encoding of local/global features and geometric properties, and *isometry-invariant* through a proper filtering of the Laplacian spectrum, *robust* to noise, and domain discretization.

Our goal is a unified review of the definition, discretization, and computation of Laplacian spectral kernels and distances, which is independent of

- the dimensionality of the input data; indeed, we will consider the case of surfaces, volumes, and n-dimensional data;

- the discretization of the input domain as a (triangle, polygonal, volumetric) mesh or a point set; and

- the discretization of the Laplace-Beltrami operator, i.e., the selection of the Laplacian weights.

After a brief overview on the Green kernel of linear operators (Sec. 3.1), we introduce the spectral operator and kernel (Sec. 3.2) by filtering the Laplacian spectrum and we discuss the stability of the computation of the Laplacian spectrum. Then, we introduce the Laplacian spectral distances (Sec. 3.3), as a generalization of the commute-time, biharmonic, diffusion, and wave distances (Sec. 3.4), and its spectrum-free computation (Sec. 3.5).

3.1   GREEN KERNEL AND LINEAR OPERATOR

Let us consider a linear differential operator $\mathcal{L}$ and the corresponding *Green kernel* $G : \mathcal{N} \times \mathcal{N} \to \mathbb{R}$ such that $\mathcal{L}G(\mathbf{p}, \mathbf{q}) = \delta(\mathbf{p} - \mathbf{q})$, where $\delta(\cdot)$ is the delta function. Then, the solution to the differential equation $\mathcal{L}u = f$ is expressed in terms of the Green kernel as

$$u(\mathbf{p}) = \int_{\mathcal{N}} G(\mathbf{p}, \mathbf{q}) f(\mathbf{q}) \mathrm{d}\mathbf{q}, \quad \mathbf{p} \in \mathcal{N}.$$

In fact,

$$(\mathcal{L}u)(\mathbf{p}) = \int_{\mathcal{N}} \mathcal{L}G(\mathbf{p}, \mathbf{q}) f(\mathbf{q}) \mathrm{d}\mathbf{q} = \int_{\mathcal{N}} \delta(\mathbf{p} - \mathbf{q}) f(\mathbf{q}) \mathrm{d}\mathbf{q} = f(\mathbf{p}).$$

Now, let us consider the integral operator $\mathcal{K}_G : \mathcal{L}_2(\mathcal{N}) \times \mathcal{L}_2(\mathcal{N}) \to \mathbb{R}$ induced by the Green kernel and defined as

$$(\mathcal{K}_G f)(\mathbf{p}) = \int_{\mathcal{N}} G(\mathbf{p}, \mathbf{q}) f(\mathbf{q}) \mathrm{d}\mathbf{q},$$

whose well posedness is guaranteed if $G \in \mathcal{L}_\infty(\mathcal{N} \times \mathcal{N})$, as a consequence of the following upper bound

$$\|\mathcal{K}_G f\|_2^2 \leq \int_{\mathcal{N}} G^2(\mathbf{p}, \mathbf{q}) \mathrm{d}\mathbf{q} \int_{\mathcal{N}} f^2(\mathbf{q}) \mathrm{d}\mathbf{q} \leq |\mathcal{N}| \|G\|_\infty^2 \|f\|_2^2,$$

where $|\cdot|$ is the area (or volume) of $\mathcal{N}$. Since $\mathcal{K}_G$ is well posed and linear, we consider its eigensystem $(\lambda_n, \phi_n)_{n=0}^{+\infty}$, $\mathcal{K}_G\phi_n = \lambda_n\phi_n$; applying $\mathcal{L}$ to both sides of this relation, we get that

$$
\begin{aligned}
\lambda_n\mathcal{L}\phi_n &= \mathcal{L}\mathcal{K}_G\phi_n \\
&= \mathcal{L}\left[\int_{\mathcal{N}} G(\cdot,\mathbf{q})\phi_n(\mathbf{q})\mathrm{d}\mathbf{q}\right] \\
&= \int_{\mathcal{N}} \mathcal{L}G(\cdot,\mathbf{q})\phi_n(\mathbf{q})\mathrm{d}\mathbf{q} \\
&= \int_{\mathcal{N}} \delta(\cdot-\mathbf{q})\phi_n(\mathbf{q})\mathrm{d}\mathbf{q} \\
&= \phi_n,
\end{aligned}
$$

i.e., $\mathcal{L}\phi_n = \lambda_n^{-1}\phi_n$, $\lambda_n \neq 0$. It follows that the eigensystem of $\mathcal{L}$ is $(\lambda_n^{-1}, \phi_n)_{n=0}^{+\infty}$; indeed, $\mathcal{L}$ and $\mathcal{K}_G$ have the same eigenfunctions and reciprocal eigenvalues.

Let us now represent the Green kernel in terms of the eigensystem of the integral operator as $G(\mathbf{p},\mathbf{q}) = \sum_{n=0}^{+\infty} \alpha_n(\mathbf{p})\phi_n(\mathbf{q})$; by definition, we have that

$$
\begin{aligned}
\delta(\mathbf{p}-\mathbf{q}) &= \mathcal{L}G(\mathbf{p},\mathbf{q}) \\
&= \mathcal{L}\left[\sum_{n=0}^{+\infty} \alpha_n(\mathbf{p})\phi_n(\mathbf{q})\right] \\
&= \sum_{n=0}^{+\infty} \alpha_n(\mathbf{p})\mathcal{L}\phi_n(\mathbf{q}) \\
&= \sum_{n=0}^{+\infty} \lambda_n^{-1}\alpha_n(\mathbf{p})\phi_n(\mathbf{q}).
\end{aligned}
$$

Multiplying both sides of the previous relation by ϕ_m, integrating over $\mathcal{N}$, and assuming that the eigenfunctions are orthonormal, we get that

$$
\begin{aligned}
\phi_m(\mathbf{p}) &= \int_{\mathcal{N}} \sum_{n=0}^{+\infty} \lambda_n^{-1}\alpha_n(\mathbf{p})\phi_n(\mathbf{q})\phi_m(\mathbf{q})\mathrm{d}\mathbf{q} \\
&= \sum_{n=0}^{+\infty} \lambda_n^{-1}\alpha_n(\mathbf{p})\langle\phi_n,\phi_m\rangle_2 \\
&= \lambda_m^{-1}\alpha_m(\mathbf{p}).
\end{aligned}
$$

Indeed, $\alpha_n(\mathbf{p}) = \lambda_n\phi_n(\mathbf{p})$ and the spectral representation of the Green kernel is

$$
G(\mathbf{p},\mathbf{q}) = \sum_{n=0}^{+\infty} \lambda_n\phi_n(\mathbf{p})\phi_n(\mathbf{q}),
$$

which is analogous to the Mercer theorem [Aro50] for the kernel of integral operator.

Let us now study the existence of the solution to the differential problem $\mathcal{L}u = f$, where $\mathcal{L}$ is a linear and self-adjoint operator. Expressing the solution $u = \sum_{n=0}^{+\infty} \alpha_n \phi_n$ in terms of the eigensystem $(\lambda_n, \phi_n)_{n=0}^{+\infty}$ of $\mathcal{L}$, we get that the relation $\lambda_n = \langle f, \phi_n \rangle_2$ must hold for $n \in \mathbb{N}$. If $\lambda_k \neq 0$, then $\alpha_k = \frac{\langle f, \phi_n \rangle_2}{\lambda_k}$. If $\lambda_k = 0$ for some k, then $\langle f, \phi_k \rangle_2 = 0$, i.e., for each null eigenvalue of $\mathcal{L}$ (if any), the right-hand side function must be orthogonal to the corresponding eigenspace $\mathcal{V}_0$. If the orthogonality condition is satisfied, then the solution is written as

$$ u = \sum_{n:\,\lambda_n \neq 0} \frac{\langle f, \phi_n \rangle_2}{\lambda_n} \phi_n + v, \quad v \in \mathcal{V}_0, $$

and it is not unique if $\mathcal{V}_0 \neq \{0\}$. If $\mathcal{V}_0$ is not trivial and the right-hand side is not orthogonal to $\mathcal{V}_0$, then there is no solution. In this last case, we can replace f with its projection $f^\perp$ in $\mathcal{V}_0^\perp$. Since the first null eigenvalue has multiplicity 1 and the corresponding eigenfunction is $\phi_0 = 1$, it is enough to consider

$$ f^\perp = f - \langle f, 1 \rangle_2 = f - \int_{\mathcal{M}} f(\mathbf{p})\mathrm{d}\mathbf{p}. $$

3.2   LAPLACIAN SPECTRAL OPERATOR AND KERNEL

Starting from recent work on the geodesic and heat diffusion distances [CWW13, Pat13], we address the definition of spectral distances on a manifold $\mathcal{N}$ by filtering its Laplacian spectrum [BB11a, Pat14]. Given a strictly positive *filter function* $\varphi : \mathbb{R}^+ \to \mathbb{R}$, let us consider the power series $\varphi(s) = \sum_{n=0}^{+\infty} \alpha_n s^n$. Noting that $\Delta^i f = \sum_{n=0}^{+\infty} \lambda_n^i \langle f, \phi_n \rangle_2 \phi_n$, we define the *spectral operator* as

$$ \Phi_\varphi(f) = \sum_{n=0}^{+\infty} \alpha_n \Delta^n f = \sum_{n=0}^{+\infty} \varphi(\lambda_n) \langle f, \phi_n \rangle_2 \phi_n. \tag{3.1} $$

The spectral operator is also linear, symmetric, positive semi-definite, and *self-adjointness*; in fact,

$$ \langle \Phi_\varphi(f), g \rangle_2 = \langle f, \Phi_\varphi(g) \rangle_2 = \sum_{n=0}^{+\infty} \varphi(\lambda_n) \langle f, \phi_n \rangle_2 \langle g, \phi_n \rangle_2. $$

Since $(\varphi(\lambda_n), \phi_n)_{n=0}^{+\infty}$ is the eigensystem of Φ_φ, we conclude that Φ_φ has a null eigenvalue if and only only if $\varphi(\lambda_l) = 0$ for some l, i.e., at least one Laplacian eigenvalue is a zero of the filter map.

We now introduce the representation and properties of the Laplacian spectral kernel (Sec. 3.2.1) and of the spectral operator (Sec. 3.2.2).

3.2.1   LAPLACIAN SPECTRAL KERNEL

Rewriting the spectral operator as

$$\Phi_\varphi(f) = \langle K_\varphi, f \rangle_2, \qquad K_\varphi(\mathbf{p}, \mathbf{q}) := \sum_{n=0}^{+\infty} \varphi(\lambda_n)\phi_n(\mathbf{p})\phi_n(\mathbf{q}), \qquad (3.2)$$

we introduce the *spectral kernel* $K_\varphi(\cdot, \cdot)$. Analogously to the diffusion kernel, the spectral kernel satisfies the following properties:

- *non-negativity*: $K_\varphi(\mathbf{p}, \mathbf{p}) \geq 0$;

- *symmetry*: $K_\varphi(\mathbf{p}, \mathbf{q}) = K_\varphi(\mathbf{q}, \mathbf{p})$; and

- *positive semi-definiteness*:

$$0 \leq \langle \Phi_\varphi(f), f \rangle_2 = \int_{\mathcal{N} \times \mathcal{N}} K_\varphi(\mathbf{p}, \mathbf{q}) f(\mathbf{p}) f(\mathbf{q}) d\mathbf{p} d\mathbf{q}$$
$$= \sum_{n=0}^{+\infty} \varphi(\lambda_n) |\langle f, \phi_n \rangle_2|^2.$$

Introducing the weighted least-squares error

$$\mathcal{E}(g) := \int_{\mathcal{N}} K_\varphi(\mathbf{p}, \mathbf{q}) |f(\mathbf{p}) - g(\mathbf{q})|^2 d\mathbf{p},$$

its minimum with respect to g is achieved by normalizing the function $\Phi_\varphi(f)$ by the area (or volume) $|\mathcal{N}|$ of the domain $\mathcal{N}$. In fact, the solution to the normal equation $0 = \partial_g \mathcal{E} = \int_{\mathcal{N}} K_\varphi(\mathbf{p}, \mathbf{q})(f(\mathbf{p}) - g(\mathbf{q})) d\mathbf{p}$ is

$$g(\mathbf{q}) = |\mathcal{N}|^{-1} \int_{\mathcal{N}} K_\varphi(\mathbf{p}, \mathbf{q}) f(\mathbf{p}) d\mathbf{p} = |\mathcal{N}|^{-1} \Phi_\varphi(f).$$

We notice that the spectral kernel is *square integrable* $\|K_\varphi\|_2^2 = \sum_{n=0}^{+\infty} |\varphi(\lambda_n)|^2$, which is equivalent to the Parseval's equality, and the *conservation property* $\int_{\mathcal{N}} K_\varphi(\mathbf{p}, \mathbf{q}) d\mathbf{p} = 1$ is a consequence of the Perron-Frobenious theorem.

We now discuss the condition on the filter that guarantees the well-posedness of the spectral kernel. According to [Hoe68, Sog88], the Laplacian eigenfunction ϕ_n on $\mathcal{N}$ associated with the eigenvalue λ_n, $\lambda_n \neq 0$, satisfies the upper bound $\|\phi_n\|_\infty \leq C \lambda_n^{1/4} \|\phi_n\|_2$, where C is a constant that depends only on the geometric properties (i.e., sectional curvature, injectivity radius) of $\mathcal{N}$. From the upper bound

$$K_\varphi(\mathbf{p}, \mathbf{q}) \leq \sum_{n=0}^{+\infty} \varphi(\lambda_n) \|\phi_n\|_\infty^2 \leq C^2 \sum_{n=0}^{+\infty} \lambda_n^{1/2} \varphi(\lambda_n),$$

we get that the integrability of $\tilde{\varphi}(s) := s^{1/2}\varphi(s)$ on $\mathbb{R}^+$ guarantees the well-posedness of the spectral kernel.

Finally, we notice that

$$\begin{cases} (f \star g)(\mathbf{p}) = \sum_{n=0}^{+\infty} \varphi(\lambda_n)\langle f, \phi_n \rangle_2 \langle g, \phi_n \rangle_2 \phi_n(\mathbf{p}), \\ (K_\varphi(\mathbf{p}, \cdot) \star f)(\mathbf{q}) = \sum_{n=0}^{+\infty} \varphi(\lambda_n)\langle f, \phi_n \rangle_2 \phi_n(\mathbf{p})\phi_n(\mathbf{q}); \end{cases}$$

in particular,

$$\int_{\mathcal{N}} \|K_\varphi(\mathbf{p}, \cdot) \star f\|_2^2 d\mathbf{p} = \sum_{n=0}^{+\infty} \varphi^2(\lambda_n)|\langle f, \phi_n \rangle_2|^2 = \|\Phi_\varphi(f)\|_2^2.$$

Green kernel of the Laplacian spectral operator   The results discussed in Sec. 3.1 can be easily specialized to the spectral operator $\mathcal{L} := \Phi_\varphi$, whose Green kernel $G(\mathbf{p}, \mathbf{q}) = \sum_{n=0}^{+\infty} \frac{1}{\varphi(\lambda_n)}\phi_n(\mathbf{p})\phi_n(\mathbf{q})$ induces the pseudo-inverse operator $\mathcal{L}^\dagger = \Phi_{1/\varphi}$, which can be interpreted as the integral operator induced by $G(\cdot, \cdot)$. We also notice that the integral operator $\mathcal{K}_G$ induced by the Green kernel and the spectral operator are commutative, i.e., $\Phi_\varphi \mathcal{K}_G = \mathcal{K}_G \Phi_\varphi$. In fact,

$$\Phi_\varphi \mathcal{K}_G f = \sum_{n=0}^{+\infty} \lambda_n \varphi(\lambda_n)\langle f, \phi_n \rangle_2 \phi_n = \mathcal{K}_G \Phi_\varphi f = \tilde{\varphi}(\Delta)f,$$

where the new filter is $\tilde{\varphi}(s) := s\varphi(s)$. If we consider the spectral operator, then the problem $\Phi_\varphi u = f$ always admits a unique solution if the filer φ is strictly positive. Otherwise, f must be orthogonal to the eigenfunctions ϕ_k such that $\varphi(\lambda_k) = 0$ (if any). If this condition is not satisfied, then we can replace f with its projection $f^\perp$ in $V_0^\perp$, $V_0 := \text{span}\{\phi_i, \lambda_i = 0\}$, which is defined as $f^\perp = f - \sum_{i, \lambda_i = 0}\langle f, \phi_i \rangle_2 \phi_i$.

3.2.2   SPECTRUM OF THE SPECTRAL OPERATOR

We show that the computation of single eigenvalues of the spectral operator is numerically stable and instabilities are generally due to repeated or close eigenvalues. First, we notice that if λ is a Laplacian eigenvalue of multiplicity m then $\mu := \varphi(\lambda)$ is an eigenvalue of Φ_φ and its multiplicity is equal to or greater than m. The corresponding eigenfunction is the Laplacian eigenfunction associated with the eigenvalue λ, i.e., $\Delta\phi = \lambda\phi$ and $\Phi_\varphi\phi = \varphi(\lambda)\phi$.

We perturb the spectral operator by $\delta\mathcal{E}$, $\delta \to 0$, and compute the eigenpair $(\mu(\delta), \phi(\delta))$ of the corresponding operator $\Phi_\varphi + \delta\mathcal{E}$, i.e.,

$$(\Phi_\varphi + \delta\mathcal{E})\,\phi(\delta) = \mu(\delta)\phi(\delta), \qquad \phi(0) = \phi, \quad \mu(0) = \mu. \tag{3.3}$$

The size of the derivative of $\mu(\delta)$ indicates the variation that it undergoes when the spectral operator is perturbed by $(\mathcal{E}, \delta)$. Deriving (3.3) with respect to δ and evaluating the resulting relation at 0, we get that

$$\mathcal{E}\phi + \Phi_\varphi \phi'(0) = \mu'(0)\phi + \mu\phi'(0),\qquad(3.4)$$

and from the self-adjointness of Φ_φ it follows that $\langle \phi, \Phi_\varphi \phi'(0)\rangle_2 = \mu \langle \phi, \phi'(0)\rangle_2$. Multiplying both sides of (3.4) by ϕ and applying the previous identity, we get that

$$|\mu'(0)| = |\langle \phi, \mathcal{E}\phi\rangle_2| \le \|\mathcal{E}\|_2 \|\phi\|_2^2 = \|\mathcal{E}\|_2.$$

We conclude that the computation of the eigenvalue of Φ_φ with multiplicity one is stable. Let us assume that $\mu := \varphi(\lambda)$ is an eigenvalue of Φ_φ with multiplicity m. Rewriting the characteristic polynomial as $p(s) = (s - \mu)^m q(s)$, where $q(\cdot)$ is a polynomial of degree $n - m$ and $q(\mu) \ne 0$, we get that $(s - \mu)^m = \mathcal{O}(\delta)/q(s)$, i.e., $s \approx \mu + \mathcal{O}(\delta^{\frac{1}{m}})$. Indeed, modifying the Laplacian matrix in such a way that the filtered eigenvalues are perturbed by $\delta := 10^{-m}$ corresponds to a change of order 0.1 in μ (i.e., $s \approx \mu + 0.1$) and this amplification becomes larger while increasing the multiplicity of the eigenvalue.

While multiple eigenvalues are typically associated with symmetric shapes, numerically close or switched eigenvalues are present regardless of the surface regularity. This unstable computation of multiple eigenpairs of the spectral operator generally affects the accuracy of the truncated spectral approximation $d_k(\mathbf{p}, \mathbf{q}) := \sum_{n=0}^{k} \frac{|\phi_n(\mathbf{p}) - \phi_n(\mathbf{q})|^2}{\varphi^2(\lambda_n)}$ of the corresponding distances. In fact, each filtered eigenvalue $\varphi(\lambda_n)$, $n = 1, \ldots, k$, which appears at the denominator of d_k, can further accentuate the numerical error of λ_n in the distance computation. Furthermore, the computation of the Laplacian spectrum is time-consuming and it is difficult to properly select the number of eigenpairs that is necessary to accurately approximate the spectral distance. Indeed, in Sec. 3.5 we propose an evaluation of the spectral distance that is independent of the computation of the Laplacian spectrum and is equivalent to a set of differential equations involving only the Laplace-Beltrami operator. This novel spectrum-free computation is a generalization of the discrete approach recently presented in [Pat16a].

3.3 LAPLACIAN SPECTRAL DISTANCES

Through the spectral operator, we introduce the scalar product and the corresponding distance as [Pat16a]

$$\begin{cases} \langle f, g\rangle := \langle \Phi_{1/\varphi} f, \Phi_{1/\varphi} g\rangle_2 = \sum_{n=0}^{+\infty} \frac{\langle f, \phi_n\rangle_2 \langle g, \phi_n\rangle_2}{\varphi^2(\lambda_n)}, & (a) \\ d^2(f, g) := \|f - g\|^2 = \sum_{n=0}^{+\infty} \frac{|\langle f - g, \phi_n\rangle_2|^2}{\varphi^2(\lambda_n)}. & (b) \end{cases} \qquad (3.5)$$

Indicating with $\delta_\mathbf{p}$ the map that takes value 1 at $\mathbf{p}$ and 0 otherwise, and selecting $f := \delta_\mathbf{p}, g := \delta_\mathbf{q}$ in Eq. (3.5b), the *spectral distance* (Fig. 3.1, Fig. 3.2) on $\mathcal{N}$ is defined as

$$
\begin{aligned}
d^2(\mathbf{p}, \mathbf{q}) &:= \|\delta_\mathbf{p} - \delta_\mathbf{q}\|^2 \\
&=_{\text{Eq. (3.5(b))}} \sum_{n=0}^{+\infty} \frac{|\phi_n(\mathbf{p}) - \phi_n(\mathbf{q})|^2}{\varphi^2(\lambda_n)} \\
&=_{\text{Eq. (3.5(a))}} \|\Phi_{1/\varphi}(\delta_\mathbf{p}) - \Phi_{1/\varphi}(\delta_\mathbf{q})\|_2^2 \\
&= \|K_{1/\varphi}(\mathbf{p}, \cdot) - K_{1/\varphi}(\mathbf{q}, \cdot)\|_2^2 \\
&= K_{1/\varphi}(\mathbf{p}, \mathbf{p}) - 2K_{1/\varphi}(\mathbf{p}, \mathbf{q}) + K_{1/\varphi}(\mathbf{q}, \mathbf{q}).
\end{aligned}
\tag{3.6}
$$

Equation (3.6) presents different formulations of the spectral distances in terms of the Laplacian spectrum, the spectral operator, and kernel. The third equality follows from the identity $\Phi_\varphi(\delta_\mathbf{p}) = K_\varphi(\mathbf{p}, \cdot)$ and it will be used for the computation of the spectral distances (Sec. 4.3.3); in fact, it is independent of the evaluation of the Laplacian spectrum. The last equality is achieved by applying the relation

$$
\begin{aligned}
d^2(\mathbf{p}, \mathbf{q}) &= \|K_{1/\varphi}(\mathbf{p}, \cdot) - K_{1/\varphi}(\mathbf{q}, \cdot)\|_2^2 \\
&= \sum_{n=0}^{+\infty} \varphi^{-2}(\lambda_n)\phi_n(\mathbf{p})\phi_n(\mathbf{p}) - 2\sum_{n=0}^{+\infty} \varphi^{-2}(\lambda_n)\phi_n(\mathbf{p})\phi_n(\mathbf{q}) + \sum_{n=0}^{+\infty} \varphi^{-2}(\lambda_n)\phi_n(\mathbf{q})\phi_n(\mathbf{q}) \\
&= K_{1/\varphi^2}(\mathbf{p}, \mathbf{p}) - 2K_{1/\varphi^2}(\mathbf{p}, \mathbf{q}) + K_{1/\varphi^2}(\mathbf{q}, \mathbf{q}).
\end{aligned}
$$

We now identify the properties of the filter that guarantees the well-posedness of the spectral kernel and distances (Sec. 3.3.1), and the scale invariance of the spectral signatures (Sec. 3.3.2).

3.3.1   WELL-POSEDNESS OF THE SPECTRAL KERNELS AND DISTANCES

According to [Hoe68, Sog88], the Laplacian eigenfunction ϕ_n on a 2-manifold $\mathcal{N}$ and associated with the eigenvalue λ_n, $\lambda_n \neq 0$, satisfies the upper bound $\|\phi_n\|_\infty \leq C\lambda_n^{1/4}\|\phi_n\|_2$, where C is a constant that depends only on the geometric properties (i.e., sectional curvature, injectivity radius) of $\mathcal{N}$. Selecting eigenfunctions with unitary norm, the spectral distance satisfies the upper bound

$$
d^2(\mathbf{p}, \mathbf{q}) \leq 4\sum_{n=0}^{+\infty} \frac{\|\phi_n\|_\infty^2}{\varphi^2(\lambda_n)} \leq 4C^2 \sum_{n=0}^{+\infty} \frac{\lambda_n^{1/2}}{\varphi^2(\lambda_n)}.
$$

If the map $\tilde{\varphi}(s) := s^{1/2}/\varphi^2(s)$ is integrable on $\mathbb{R}^+$, then the series that defines the spectral distance is convergent. The nullity condition $d(\mathbf{p}, \mathbf{q}) = 0$ holds if and only if $\phi_i(\mathbf{p}) = \phi_i(\mathbf{q})$, $\forall i \in \mathbb{N}$, i.e., $\langle 1_\mathbf{p} - 1_\mathbf{q}, \phi_i \rangle_2 = 0$, where $1_\mathbf{p}(\cdot)$ has value one at $\mathbf{p}$ and zero otherwise. Noting that $\{\phi_i\}_{i=0}^{+\infty}$ is a basis of $\mathcal{L}_2(\mathcal{N})$, the nullity relation is satisfied if and only if $(1_\mathbf{p} - 1_\mathbf{q})$ is the null function, i.e., $\mathbf{p} = \mathbf{q}$. The symmetry and triangular inequality follow from Eq. (3.6).

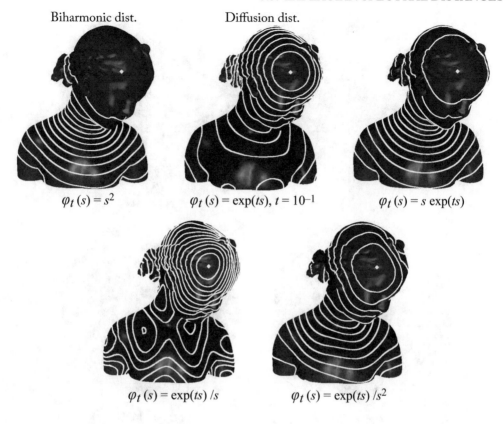

Figure 3.1: Level sets of the spectral distances from a source point (white dot) induced by the filter φ and evaluated with the Padé-Chebyshev approximation ($r = 5$).

According to the results in Sec. 3.2, the integrability of $\tilde{\varphi}(s) := s^{1/2}/\varphi(s)$ on $\mathbb{R}^+$ guarantees the well-posedness of the spectral kernel. In particular, if the Laplacian eigenfunctions are uniformly bounded (i.e., $\|\phi_n\|_\infty \leq C_\infty$, $\forall n \in \mathbb{N}$) then the integrability of $1/\varphi$ guarantees the well-posedness of the spectral kernel and distance. This hypothesis applies only to the continuous case; in fact, their discrete counterparts (c.f., Eqs. (4.1), (4.2)) are defined through a finite sum.

3.3.2   SCALE INVARIANCE AND SHAPE SIGNATURES

Rescaling $\mathcal{N}$ to $\alpha\mathcal{N}$, the eigenpairs of $\alpha\mathcal{N}$ becomes $(\alpha^{-2}\lambda_n, \alpha^{-1}\phi_n)_{n=0}^{+\infty}$ and the corresponding distances becomes

$$d_{\alpha\mathcal{N}}(\mathbf{p}, \mathbf{q}) = \sum_{n=0}^{+\infty} \alpha^{-2}\varphi^{-2}(\alpha^{-2}\lambda_n)|\phi_n(\mathbf{p}) - \phi_n(\mathbf{q})|^2.$$

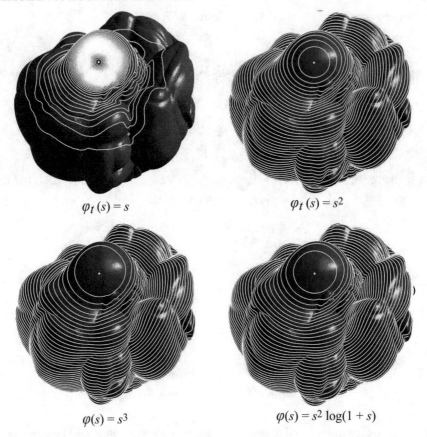

$$\varphi_t(s) = s \qquad\qquad \varphi_t(s) = s^2$$

$$\varphi(s) = s^3 \qquad\qquad \varphi(s) = s^2 \log(1+s)$$

Figure 3.2: Level sets of the spectral distances from a source point (white dot) induced by the filter φ and evaluated with the Padé-Chebyshev approximation ($r = 5$).

Indeed, $d_{\alpha\mathcal{N}} = d_{\mathcal{N}}$ if and only if $\varphi(s) = \alpha\varphi(\alpha^{-2}s)$, $\alpha > 0$. For instance, the filter $\varphi(s) := s^{1/2}$ associated with the commute-time distances satisfies the previous relation. For the spectral kernel, we note that

$$K_\varphi^{\alpha\mathcal{N}}(\mathbf{p}, \mathbf{q}) = \sum_{n=0}^{+\infty} \alpha^{-2}\varphi(\alpha^{-2}\lambda_n)\phi_n(\mathbf{p})\phi_n(\mathbf{q})$$

and the spectral kernel is invariant to scale deformation if and only if $\alpha^{-2}\varphi(\alpha^{-2}s) = \varphi(s)$, $\forall s$. A similar case applies to the discrete Laplacian spectral kernels and distances. Equivalent representations of the Laplacian spectral distances are summarized in Table 3.1.

Through the selected filter function and the Laplacian spectrum, we define the *spectral embedding* $\mathcal{E} : \mathcal{M} \to \ell_2$, which maps each point $\mathbf{p}$ to the sequence $\mathcal{E}(\mathbf{p}) := (\varphi(\lambda_n)\phi_n(\mathbf{p}))_{n=0}^{+\infty}$. The equality $d(\mathbf{p}, \mathbf{q}) = \|\mathcal{E}(\mathbf{p}) - \mathcal{E}(\mathbf{q})\|_2$ shows that the spectral distances can be inter-

Table 3.1: Equivalent representations of the Laplacian spectral distances

Spectral Distance Representation $d(\mathbf{p}, \mathbf{q})$	Involved Terms	Details
$\sum_{n=0}^{+\infty} \varphi^{-2}(\lambda_n)\|\phi_n(\mathbf{p}) - \phi_n(\mathbf{q})\|^2$	Lapl. spectrum	Eq. (4.6) $(\lambda_n, \phi_n)_{n=0}^{+\infty} : \Delta \phi_n = \lambda_n \phi_n$
$\| K_{1/\varphi}(\mathbf{p},\cdot) - K_{1/\varphi}(\mathbf{q},\cdot) \|_2^2$	Lapl. kernel	Eq. (4.6)
$K_{1/\varphi^2}(\mathbf{p}, \mathbf{p}) - 2K_{1/\varphi^2}(\mathbf{p}, \mathbf{q}) + K_{1/\varphi^2}(\mathbf{p}, \mathbf{p})$	Lapl. kernel	Eq. (4.3)
$\| \Phi_{1/\varphi}(\delta_\mathbf{q}) - \Phi_{1/\varphi}(\delta_\mathbf{q}) \|_2^2$	Lapl. spectral oper.	$\Phi_{1/\varphi}(\delta_\mathbf{q}) = K_{1/\varphi}(\mathbf{p},\cdot)$
$\| \phi_n(\mathbf{p}) - \phi(\mathbf{q}) \|_2^2$	Lapl. spect. embed.	$\phi : N \to \ell_2$ $\phi(\mathbf{p}) := (\varphi^{-1}(\lambda_n)\|\phi_n(\mathbf{p}))_{n=0}^{+\infty}$

preted as Euclidean distances in the embedding space. Finally, the *spectral shape descriptor* $SD(\mathbf{p}) := \sum_{n=0}^{+\infty} |\varphi(\lambda_n)|^2\phi_n^2(\mathbf{p})$ and *signature* $SE(\mathbf{p}) := (\varphi^{-1/2}(\lambda_n)\phi_n(\mathbf{p}))_{n=0}^{+\infty}$ generalize the diffusion descriptor and signature [DLL+10, SOG09] (Sec. 3.4.2).

3.4   MAIN EXAMPLES OF SPECTRAL DISTANCES

We introduce the main criteria for the selection of the filter map (Sec. 3.4.1) and as special cases we consider the diffusion (Sec. 3.4.2), commute-time and biharmonic distances (Sec. 3.4.3), and the approximation of the geodesic and transportation distances (Sec. 3.4.4). For spectral distances, the minimum and the maximum values are depicted in blue and red, respectively. For all the other functions, colors begin with red, pass through yellow, green, cyan, blue, and magenta, and return to red. Finally, the color coding represents the same scale of values for multiple shapes.

3.4.1   SELECTION OF THE FILTER MAP

The filter is selected according to different properties of the resulting distances, such as isometry invariance, localization in both space and frequency [HVG11], or through a supervised learning [ABBK11]. The selection of φ provides a simple way to design new distances, whose smoothness and balance between the measure of both local and global properties depend on the convergence of the filtered eigenvalue $1/\varphi(\lambda_i)$ to zero or on their periodic behavior. Increasing the decay of $1/\varphi$ to zero, the effects of the larger Laplacian eigenvalues and of the corresponding eigenvectors on the spectral distance are negligible with respect to the contribution of the lower Laplacian eigenvalues. The resulting distance encodes the global shape properties. Reducing the filter growth, local shape features are better characterized. The most difficult part in the design of the filter is the selection of its behavior in the frequency domain with respect to the geometric features of the input shape that will be encoded in the corresponding spectral distances. In fact, the quantitative relation between the Laplacian eigenvalues and the frequency of the local and global features of the input domain are unknown for an arbitrary 3D shape.

For instance (Fig. 3.3), selecting $\varphi_t(s) := \exp(-st)$, $\exp(-ist)$ or $\varphi(s) := s^{-k/2}, s^{-1/2}$, we get the *heat diffusion, wave,* or *poly-harmonic, commute-time distances,* respectively. Mexican hat wavelets [HQ12] are generated by the filter $\varphi(s) := s^{1/2}\exp(-s^2)$ and in [BB11a, ASC11] the filter function $\varphi(s) := \exp(is)$, $s \in [0, 2\pi]$, defines the wave kernel signature. The spectral distances associated with this periodic filter identify local shape features by separating the contribution of different frequencies and of the corresponding eigenfunctions. We also notice that filters defined for Laplacian spectral smoothing (Fig. 3.4a,b), such as $(1 + \beta_1 s + \beta_2 s^2)^n$ [Tau95], $(1 + \alpha_1 s)^{-2}$ [DMSB99], and $(1 + \alpha_2 s^2)^{-1}$ [ZF03], are also useful for the definition of spectral distances. Since the filter in [Tau95] does not decay to zero, it is typically applied to the Laplacian matrix with constant weights, whose eigenvalues belong to $[0, 2]$.

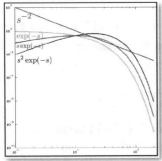

φ^2	Spect. dist.	$d^2(\mathbf{p},\mathbf{q}) = \sum_{n=0}^{+\infty} \varphi^2(\lambda_n)|\phi_n(\mathbf{p}) - \phi_n(\mathbf{q})|^2$	
		Associated equation	**Kernel**
s^{-1}	Comm.-time	$(\partial_t + \Delta^r)F(\cdot,t) = 0$	$\sum_{n=0}^{+\infty} t^k e^{-\lambda_n t}|\phi_n(\mathbf{p}) - \phi_n(\mathbf{q})|^2$
s^{-2}	Biharm.	*Poly-harm. eq.*	
s^{-k}	Poly-harm.		
e^{-st}	Heat diff.	$(\partial_t + \Delta)F(\cdot,t) = 0$	$\sum_{n=0}^{+\infty} e^{-\lambda_n t}|\phi_n(\mathbf{p}) - \phi_n(\mathbf{q})|^2$
		Heat diffusion eq.	
e^{-ist}	Wave ker.	$(\partial_t + i\Delta)F(\cdot,t) = 0$	$\sum_{n=0}^{+\infty} e^{-i\lambda_n t}|\phi_n(\mathbf{p}) - \phi_n(\mathbf{q})|^2$
		Schroedinger eq.	

Figure 3.3: Spectral distances and kernels induced by the filter function φ (log-scale on the t- and y-axis) applied to the Laplacian eigenvalues.

Similarly to random walks [FPS05, RS13], we consider the filter map $\varphi_t(s) = t^{-k}s^{-\alpha}\exp(ts^\alpha)$, where k scales the diffusion rate and α controls the distance smoothness (Fig. 3.2). The selection of the parameters α, k makes the multi-scale kernels more robust to geometric and topological noise; the integral over time also avoids the selection of the heat diffusion rate. The filter functions $\varphi_t(s) := [\cos^{-1/2}(\sqrt{st}), s^{-1/4}\sin^{1/2}(\sqrt{st})]$ and $\varphi(s,t) = \exp(s^r t)$ are associated with the diffusion equations $(\partial_t^2 + \Delta)F(\cdot,t) = 0$ and $(\partial_t + \Delta^r)F(\cdot,t) = 0$, respectively. Finally, the filter function can be learned from a set of retrieval examples [ABBK11, BMM+15]. The filters $\varphi_t(s) := [\cos^{-1}(\sqrt{st}), s^{-1/2}\sin(\sqrt{st})]$ and $\varphi(s,t) = \exp(s^r t)$ are associated with the diffusion equations $(\partial_t^2 + \Delta)F(\cdot,t) = 0$ and $(\partial_t + \Delta^r)F(\cdot,t) = 0$, respectively.

To achieve localization of the signal content in both space and frequency [HVG11], we select a filter map φ that has a power growth $\varphi(s) = s_1^{-a}s^a$ on the interval $[0, s_1)$ and a power decay $\varphi(s) = s_2^b s^{-b}$ in $(s_2, +\infty)$, $s_1 < s_2$, for large s. In $[s_1, s_2]$, φ is a cubic polynomial such that the filter map and its first derivative are continuous on $\mathbb{R}^+$. A stronger or weaker encoding

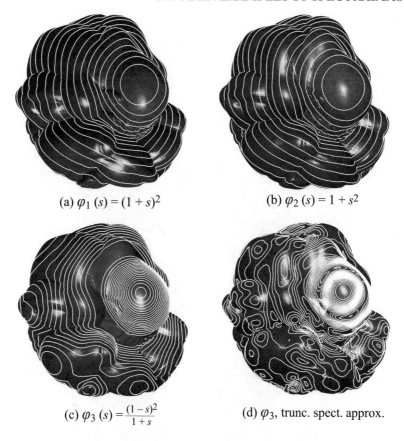

(a) $\varphi_1(s) = (1+s)^2$ (b) $\varphi_2(s) = 1 + s^2$

(c) $\varphi_3(s) = \dfrac{(1-s)^2}{1+s}$ (d) φ_3, trunc. spect. approx.

Figure 3.4: Color map and level sets of spectral distances induced by different filters and computed with (a-c) the proposed approach and (d) the truncated spectral approximation ($k = 500$ eigenpairs).

of surface details is achieved by enlarging or reducing the interval size. A common choice is $a = b = 1$, $s_1 = 1$, $s_2 = 2$, $\varphi(s) = -5 + 11s - 6s^2 + s^3$ (Fig. 3.5a,b).

In [ABBK11], a spline filter is defined through a supervised learning process that discriminates among shapes of a certain class and is insensitive to a selected family of transformations. Finally, for the definition of new distances we can consider a convex combination of the filters (Fig. 3.5(c-f)).

3.4.2 DIFFUSION DISTANCES

The heat kernel induces the *diffusion distances*, whose spectral representation is $d_t^2(\mathbf{p}, \mathbf{q}) = \sum_{n=0}^{+\infty} \exp(-\lambda_n t)|\phi_n(\mathbf{p}) - \phi_n(\mathbf{q})|^2$. Through the heat kernel, a shape is associated with a diffusion metric that measures the rate of connectivity among its points with

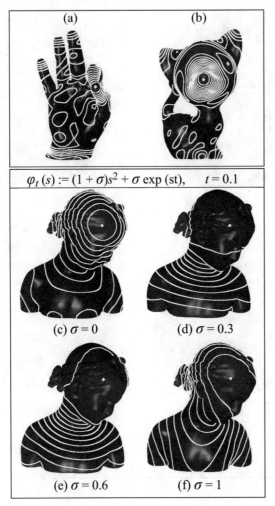

(a)   (b)

$$\varphi_t(s) := (1+\sigma)s^2 + \sigma \exp(st), \qquad t = 0.1$$

(c) $\sigma = 0$ (d) $\sigma = 0.3$

(e) $\sigma = 0.6$ (f) $\sigma = 1$

Figure 3.5: Spectral distances induced by: (a,b) the filter in [HVG11] ($r = 5$); (c-f) a convex combination of bi-harmonic ($\sigma = 0$) and diffusion ($\sigma = 1$, $t = 0.1$) distances ($r = 7$).

paths of length t and characterizes the local/global geometric behavior with small/large values of t. This property has been used to define a multi-scale and isometry-invariant signatures [BBK+10, BK10, BBOG11, CL06, DRW10, GBAL09, LKC06, MS05, Mem09, Mem11, OMMG10, RBBK10, Rus07, MS09, SOG09] and to rewrite the shape similarity problem as the comparison of two metric spaces. Main examples include the

- *heat kernel signature* $HKS(\mathbf{p}) := \sum_{n=0}^{+\infty} \exp(-\lambda_n t)|\phi_n(\mathbf{p})|^2$;

- *heat kernel descriptor* $HKD(\mathbf{p}) := (\lambda_n^{-1/2}\phi_n(\mathbf{p}))_{n=0}^{+\infty}$; and

- *wave kernel signature* $WKS(\mathbf{p}) := \sum_{n=0}^{+\infty} \exp(-i\lambda_n t)|\phi_n(\mathbf{p})|^2$.

Furthermore, the heat diffusion distance and kernel have been successfully applied to shape segmentation [dGGV08]; the computation of the gradient of discrete functions [LSW09]; and the multi-scale approximation of functions [PF10].

The diffusion distance and kernel also play a central role in several applications, such as dimensionality reduction with spectral embeddings [BN03, XHW10]; data visualization [HAvL05, RS00, TSL00], representation [CWS03, SK03, ZGL03], and classification [NJW01, SM00, ST07]. Figure 3.6 shows the stability of diffusion distances with respect to noise; in this example, the distances have been computed with the Padè-Chebyshev approximation (Sec. 4.3.2).

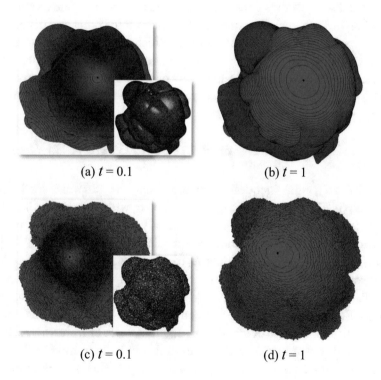

(a) $t = 0.1$ (b) $t = 1$

(c) $t = 0.1$ (d) $t = 1$

Figure 3.6: Level sets of the linear FEM diffusion distance, computed using the Padé-Chebyshev approximation ($r := 7$), from a source point (black dot), with different values of t, on a (a,b) smooth and (c,d) noisy surface.

3.4.3   COMMUTE-TIME AND BIHARMONIC DISTANCES

Integrating the diffusion distances with respect to t, we get the *commute-time distance*

$$d^2(\mathbf{p}, \mathbf{q}) = \frac{1}{2} \int_0^{+\infty} d_t^2(\mathbf{p}, \mathbf{q}) dt = \sum_{n=0}^{+\infty} \lambda_n^{-1} |\phi_n(\mathbf{p}) - \phi_n(\mathbf{q})|^2,$$

which is induced by the filter $\varphi(s) := s^{1/2}$ and is scale-invariant. We notice that this series can also be unbounded. While the diffusion distance estimates the connection of two points with respect to any random walk of length t, the commute-time distance measures this connection with respect to arbitrary random walks.

The *biharmonic distances* [OBCS+12, LRF10, Rus11b] are induced by $\varphi(s) := s$ and provide a trade-off between a nearly-geodesic behavior for small distances and global shape-awareness for large distances, thus guaranteeing an intrinsic multi-scale characterization of the input shape. In Fig. 3.7, the approximation of the biharmonic kernel and distance with a subset of the Laplacian spectrum presents local artifacts, which are represented by isolated level sets and are reduced by increasing the number of eigenpairs without disappearing. In Fig. 3.8, the smooth and uniform distribution of the level sets of the biharmonic distance around the anchor point (black dot) confirms the stability of the spectrum-free approximation with respect to surface sampling, noise, and missing parts.

3.4.4   GEODESIC AND TRANSPORTATION DISTANCES VIA HEAT KERNEL

According to [Var67], the geodesic distance can be approximated as $d_G(\mathbf{p}, \mathbf{q}) = -\lim_{t \to 0}(4t \log K_t(\mathbf{p}, \mathbf{q}))$, where $K_t(\mathbf{p}, \mathbf{q})$ is computed with the Padé-Chebyshev method. In fact, the truncated spectral approximation generally does not provide a value of $K_t(\mathbf{p}, \mathbf{q})$, $t \to 0$, enough accurate to apply the Varadhan formula. Alternatively [CWW13], the geodesic distance $d_G(\mathbf{p}, \mathbf{q})$ from the heat kernel values $K_t(\mathbf{p}, \mathbf{q})$ as the scale tends to zero. More precisely, the geodesic distance d_G on $\mathcal{N}$ is approximated by computing the solution $F(\cdot, t)$ to the heat equation $(\partial_t + \Delta)F(\cdot, t) = 0$ on $\mathcal{N}$, as $t \to 0^+$, normalizing the corresponding gradient $X = \nabla F(\cdot, t)/\|\nabla F(\cdot, t)\|_2$, and solving the equation $\Delta d_G = \operatorname{div}(X)$. This approximation is computationally efficient, capable of identifying different types of features by selecting different diffusion models, and robust to noise. The main difficulty is the tuning of the time scale with respect to the shape features; in fact, the selection of a large scale is generally associated with an over-smoothing of the geodesic values and local details.

In [SdGP+15], the *optimal transportation distances* have been approximated using the iterative Sinkhorn's method [Sin64] and the entropic regularization, thus reducing their computation to the solution of two sparse matrix equations that involve the heat kernel matrix. Instead of approximating the heat kernel with an implicit Euler integration [DMSB99], we can apply the Padé-Chebyshev approximation (Sec. 2.4.3) in order to improve the approximation accuracy at small scales and without modifying the overall approach.

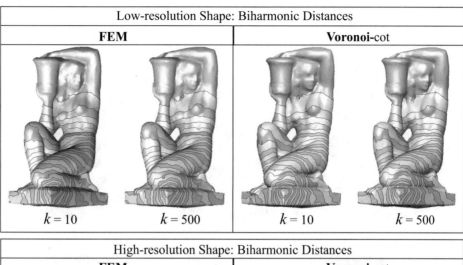

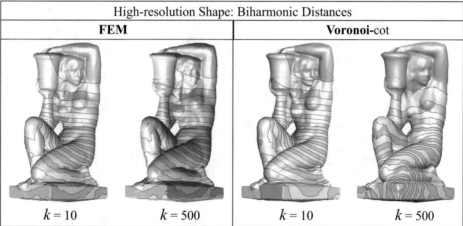

Figure 3.7: Biharmonic distance on a surface at different resolutions, with different Laplacian weights and k eigenpairs.

3.5   SPECTRUM-FREE APPROXIMATION

To characterize the spectral distances in terms of the spectral operator, we show that the pseudo-inverse of Φ_φ is induced by the filter function $1/\varphi$, i.e., $\Phi_\varphi^\dagger = \Phi_{1/\varphi}$. In fact, from the spectral representation (3.1) of $\Phi_{1/\varphi}$ and Φ_φ, we get that

$$\Phi_\varphi \Phi_{1/\varphi} \Phi_\varphi = \Phi_\varphi, \qquad \Phi_{1/\varphi} \Phi_\varphi \Phi_{1/\varphi} = \Phi_{1/\varphi}, \qquad \langle \Phi_{1/\varphi} \Phi_\varphi f, g \rangle_2 = \langle f, \Phi_{1/\varphi} \Phi_\varphi g \rangle_2.$$

Noting that

$$d(f, g) = \|u\|_2, \quad u = \Phi_{1/\varphi}(f - g) \Longleftrightarrow \Phi_\varphi u = f - g,$$

the spectral distance is equivalent to the

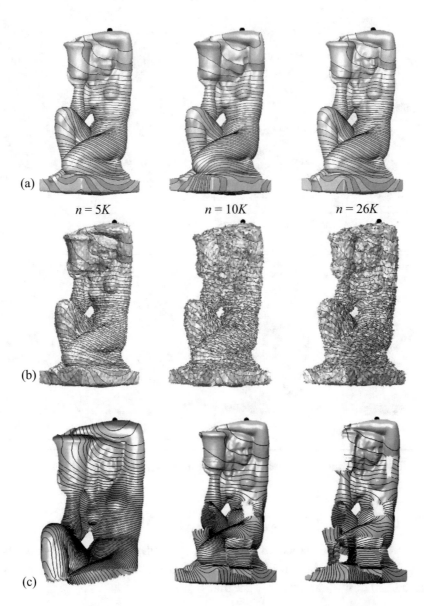

Figure 3.8: Stability of the biharmonic distance from a source (black) point with respect to (a) sampling, (b) noise, and (c) holes.

- norm of the solution of the linear equation $\Phi_{1/\varphi} u = f - g$ and

- norm of the function $u = \Phi_\varphi (f - g)$.

Through these relations, the spectral distances will be computed by approximating the filter with a polynomial or rational function and converting the evaluation of the distance to the solution of a set of differential equations that involve the Laplace-Beltrami operator (c.f., Eqs. (3.7), (3.8), (3.9)).

The selection of one of these two equivalent representations depends on the selected filter and the simplicity of evaluating either Φ_φ or $\Phi_{1/\varphi}$. In the following, we assume that both Φ_φ and $\Phi_{1/\varphi}$ are well-defined; for instance, this hypothesis is satisfied if φ is not null only on a compact interval of $\mathbb{R}^+$ and it is valid in the discrete case (Chapter 4), where the Laplacian spectrum belongs to the interval $\mathcal{I} := [0, \lambda_{\max}(\tilde{\mathbf{L}})]$, with $\lambda_{\max}(\tilde{\mathbf{L}})$ maximum Laplacian eigenvalue.

Let us now introduce the space of the Laplacian spectral kernels as

$$\mathcal{K}(\mathcal{N}) := \{K_\varphi : \mathcal{N} \times \mathcal{N} \to \mathbb{R}, \; K_\varphi \text{ spectral kernel in Eq. (3.2)}\},$$

and equipped with the $\mathcal{L}_2(\mathcal{N} \times \mathcal{N})$ scalar product. Since

$$\|K_\rho - K_\varphi\|_2^2 = \sum_{n=0}^{+\infty} |\rho(\lambda_n) - \varphi(\lambda_n)|^2 \leq \|\rho - \varphi\|_2^2.$$

the approximation of a given spectral kernel K_φ with a new kernel K_ρ in $\mathcal{K}(\mathcal{N})$ is reduced to compute the approximation of φ by ρ on a proper subspace of functions (e.g., the space of polynomials or rational polynomials). The class of functions used for the approximation of the input filter φ is selected in such a way that K_ρ provides a good approximation of the input kernel K_φ and is easily computable. To this end, we distinguish the following cases: polynomial filter (Sec. 3.5.1), polynomial approximation of the filter (Sec. 3.5.2), rational approximation of the filter with (Sec. 3.5.3) or without factorization (Sec. 3.5.4). Finally, we discuss the convergence and accuracy of the spectrum-free approximation of the spectral distances (Sec. 3.5.5).

3.5.1 POLYNOMIAL FILTER

If $\varphi(s) := \sum_{i=0}^{r} \alpha_i s^i$ is a *polynomial filter* (e.g., commute-time, poly-harmonic distances), then the distance $d(f, g) = \|u\|_2$ is equal to the norm of the solution to the equation $\Phi_\varphi u = f - g$, where Φ_φ is computed from the representation of φ and Δ. As main examples, we mention the commute-time (i.e., $\varphi(s) = s$) and bi-harmonic (i.e., $\varphi(s) = s^2$) operators.

3.5.2 ARBITRARY FILTER: POLYNOMIAL APPROXIMATION

If φ is an *arbitrary filter* (e.g., diffusion, wave kernel distances), then we approximate $1/\varphi$ with a polynomial p_r of degree r in such a way that $\Phi_{1/\varphi}$ is well approximated by Φ_{p_r} and the function $\Phi_{p_r}(f - g)$ is easily computable. In this case, the evaluation of the corresponding distance reduces to the solution of r differential equations involving the Laplace-Beltrami operator only. More precisely, computing the best r-degree polynomial approximation $p_r(s) = \sum_{i=0}^{r} \alpha_i s^i$ of $1/\varphi$ with respect to the $\mathcal{L}_\infty$-norm, we get that

$$u = \Phi_{1/\varphi}(f - g) \approx p_r(\Delta)(f - g) = \sum_{i=0}^{r} \alpha_i \Delta^i (f - g). \tag{3.7}$$

3.5.3   ARBITRARY FILTER: RATIONAL APPROXIMATION

Alternatively, we apply the *Padé-Chebyshev rational approximation* to the filter map $1/\varphi$. Let $\mathcal{R}_r^l$ be the space of all rational functions

$$c_{rl}(s) := \frac{p_r^l(s)}{q_r^l(s)} = \frac{\beta_0 + \beta_1 s + \ldots + \beta_l s^l}{\alpha_0 + \alpha_1 s + \ldots + \alpha_r s^r}.$$

We briefly recall that the representation of the rational approximation is not unique, unless we impose that it is irreducible; for example, by choosing $q_r^l(a) = 1$. Given a filter $\varphi : [a, b] \to \mathbb{R}$, defined on the finite real interval $[a, b]$, there exists a unique best approximation of $1/\varphi$ in $\mathcal{R}_r^l$ with respect to the ℓ_∞ norm. Selecting $l = r$, we have

$$\Phi_{1/\varphi}(\Delta)f \approx \frac{p_r(\Delta)f}{q_r(\Delta)} = g \iff q_r(\Delta)g = p_r(\Delta)f$$

$$\iff \sum_{i=0}^{r} \alpha_i \Delta^i g = p_r(\Delta)f, \tag{3.8}$$

i.e., g is computed by solving a new differential equation that involves the Laplace-Beltrami operator and its powers. If $\alpha_0 = 1$, $\alpha_i = 0$, $i = 1, \ldots, r$, then the rational approximation reduces to the evaluation of the function $g = p_r(\Delta)f$ by applying the differential operator $p_r(\Delta) = \sum_{i=0}^{l} \beta_i \Delta^i$ to the function f.

3.5.4   ARBITRARY FILTER: FACTORIZATION OF THE RATIONAL APPROXIMATION

For specific filters (e.g., exponential filter of the diffusion kernel/distance), the best (r, r)-degree rational polynomial approximation of $1/\varphi$ on $\mathbb{R}^+$ with respect to the $\mathcal{L}_\infty$ norm can be written as $c_{rr}(x) = \alpha_0 + \sum_{i=1}^{r} \alpha_i (1 + \beta_i x)^{-1}$, where $(\alpha_i)_{i=1}^{r}$ are the weights and $(\beta_i)_{i=1}^{r}$ are the nodes of the r-point Gauss-Legendre quadrature rule [GV89, Ch. 11]. Both the poles and coefficients are precomputed for any degree [CRV84]. Indicating with id($\cdot$) the identity operator, $\Phi_{1/\varphi}f$ is approximated as a linear combination of the solutions to the following r equations:

$$\Phi_{1/\varphi} f \approx \sum_{i=1}^{r} \alpha_i (\mathrm{id} + \beta_i \Delta)^{-1} f$$

$$= \alpha_0 f + \sum_{i=1}^{r} \alpha_i g_i, \quad (\mathrm{id} + \beta_i \Delta) g_i = f. \tag{3.9}$$

Indeed, we compute a new basis $\mathcal{B} := \{g_i\}_{i=1}^{r}$, which replaces the Laplacian eigenfunctions and is induced by the input domain, the selected filter, or its approximation, the Laplace-Beltrami operator, and f. To compute the solution to Eq. (3.9), we notice that the sequence $(g_k)_{k=0}^{+\infty}$ defined by the relation $g_{k+1} = -\beta_i \Delta g_k + f$, $g_0 := f$, converges to the solution to equation $(\mathrm{id} + \beta_i \Delta) g = f$.

3.5.5 CONVERGENCE AND ACCURACY

To verify that the sequence

$$(\Phi_{1/\varphi}^{(r)} f)_{r=0}^{+\infty}, \quad \Phi_{1/\varphi}^{(r)} f := \sum_{n=0}^{+\infty} c_{rl}(\lambda_n) \langle f, \phi_n \rangle_2 \phi_n,$$

induced by the Padé-Chebyshev rational polynomial, converges to $\Phi_{1/\varphi} f$, we apply the upper bound

$$\left\| \Phi_{1/\varphi}^{(r)} f - \Phi_{1/\varphi} f \right\|_2^2 \leq \|c_{rl} - 1/\varphi\|_\infty^2 \sum_{n=0}^{+\infty} |\langle f, \phi_n \rangle_2|^2$$

$$= \sigma_{rl}^2 \|f\|_2^2, \quad \sigma_{rl} \approx \mathcal{O}(s^{r+l+1}), \quad s \to 0,$$

where σ_{rl} is the approximation error between $1/\varphi$ and c_{rl}. For instance, we recall that the $\mathcal{L}_\infty$ error between the exponential filter (i.e., diffusion distances) and its rational polynomial approximation is bounded by the uniform rational Chebyshev constant σ_{rr}, which is known, independent of the evaluation point, and lower than 10^{-r} [Var90].

CHAPTER 4

Discrete Spectral Distances

We introduce the discrete spectral kernels and distances (Sec. 4.1), together with the induced native spaces (Sec. 4.2). Then, we discuss the computation of the spectral distances (Sec. 4.3) and main examples (Sec. 4.4).

4.1 DISCRETE SPECTRAL KERNELS AND DISTANCES

The spectral operator $\Phi_{1/\varphi}$ is discretized by the *spectral kernel matrix* $\mathbf{K}_{1/\varphi}$. Rewriting Eq. (3.2) as $\langle K_{1/\varphi}(\mathbf{p},\cdot),\phi_l\rangle_2 = \langle\phi_l,1_\mathbf{p}\rangle_2/\varphi(\lambda_l)$ and sampling it on $\mathcal{P}$, the weak formulation $\langle K_{1/\varphi}(\mathbf{p}_i,\cdot),\phi_l\rangle_2 = \langle\phi_l,\mathbf{e}_i\rangle_2/\varphi(\lambda_l)$ of the kernel is equivalent to

$$\langle\mathbf{K}_{1/\varphi}\mathbf{e}_i,\mathbf{x}_l\rangle_\mathbf{B} = \frac{\langle\mathbf{x}_l,\mathbf{e}_i\rangle_\mathbf{B}}{\varphi(\lambda_l)} \iff \mathbf{K}_{1/\varphi}\mathbf{e}_i = \sum_{l=1}^n \frac{\langle\mathbf{x}_l,\mathbf{e}_i\rangle_\mathbf{B}}{\varphi(\lambda_l)}\mathbf{x}_l,$$

where $\langle\mathbf{f},\mathbf{g}\rangle_\mathbf{B} := \mathbf{f}^\top\mathbf{B}\mathbf{g}$ is the inner product induced by $\mathbf{B}$. Indeed, the *filtered kernel matrix* is $\mathbf{K}_{1/\varphi} = \mathbf{X}\varphi^\dagger(\Lambda)\mathbf{X}^\top\mathbf{B}$, $\varphi^\dagger(\Lambda) := \text{diag}\,(1/\varphi(\lambda_i))_{i=1}^n$, and the *spectral distances* are

$$d^2(\mathbf{p}_i,\mathbf{p}_j) = \|\mathbf{K}_{1/\varphi}(\mathbf{e}_i-\mathbf{e}_j)\|_\mathbf{B}^2 = \sum_{l=1}^n \frac{|\langle\mathbf{x}_l,\mathbf{e}_i-\mathbf{e}_j\rangle_\mathbf{B}|^2}{\varphi^2(\lambda_l)}. \tag{4.1}$$

If $\varphi(\lambda_i) = 0$, then the corresponding entry in $\varphi^\dagger(\Lambda)$ is chosen equal to zero. In the last equality of Eq. (4.1), we have applied the identity $\mathbf{K}_{1/\varphi}^\top\mathbf{B}\mathbf{K}_{1/\varphi} = \mathbf{B}\mathbf{X}[\varphi^\dagger(\Lambda)]^2\mathbf{X}^\top\mathbf{B}$. If $\varphi(\lambda_i) = 0$, then the corresponding entry of the matrix $\varphi^\dagger(\Lambda) = \text{diag}(1/\varphi(\lambda_i))_{i=1}^n$ is chosen equal to zero [Pat16a].

Writing the filter map as the power series $\varphi(s) = \sum_{n=0}^{+\infty}\alpha_n s^n$ and noting that $\tilde{\mathbf{L}} = \mathbf{X}\Lambda\mathbf{X}^\top\mathbf{B}$, we get that

$$\varphi(\tilde{\mathbf{L}}) = \mathbf{X}\sum_{n=0}^{+\infty}\alpha_n\Lambda^n\mathbf{X}^\top\mathbf{B} = \mathbf{X}\varphi(\Lambda)\mathbf{X}^\top\mathbf{B} = \mathbf{K}_\varphi, \tag{4.2}$$

i.e., the spectral kernel $\mathbf{K}_\varphi = \varphi(\tilde{\mathbf{L}})$ is a filtered version of the Laplacian matrix. Furthermore, $\mathbf{K}_\varphi$ is the pseudo-inverse of $\mathbf{K}_{1/\varphi}$, i.e., $\mathbf{K}_\varphi\mathbf{K}_{1/\varphi}\mathbf{K}_\varphi = \mathbf{K}_\varphi$, $\mathbf{K}_{1/\varphi}\mathbf{K}_\varphi\mathbf{K}_{1/\varphi} = \mathbf{K}_{1/\varphi}$, and the matrices $\mathbf{K}_\varphi\mathbf{K}_{1/\varphi}$, $\mathbf{K}_{1/\varphi}\mathbf{K}_\varphi$ are $\mathbf{B}$-adjoint. In fact,

$$\mathbf{K}_\varphi\mathbf{K}_{1/\varphi}\mathbf{K}_\varphi = \mathbf{X}\varphi(\Lambda)\varphi^\dagger(\Lambda)\varphi(\Lambda)\mathbf{X}^\top\mathbf{B} = \mathbf{X}\varphi(\Lambda)\mathbf{X}^\top\mathbf{B} = \mathbf{K}_\varphi,$$
$$(\mathbf{K}_\varphi\mathbf{K}_{1/\varphi})^\top\mathbf{B} = \mathbf{B}\mathbf{X}\varphi^\dagger(\Lambda)\varphi(\Lambda)\mathbf{X}\mathbf{B} = \mathbf{B}(\mathbf{K}_\varphi\mathbf{K}_{1/\varphi}).$$

In previous work, the spectral distances have been discretized as $d(\mathbf{p}_i, \mathbf{p}_j) = \|\mathbf{K}_\varphi^\dagger(\mathbf{e}_i - \mathbf{e}_j)\|_2$ and with respect to the Euclidean scalar product, where $\mathbf{K}_\varphi^\dagger := \mathbf{X}\varphi(\Lambda)\mathbf{X}^\top$ is the corresponding kernel.

This last discretization does not take into account the intrinsic $\mathbf{B}$-scalar product, thus disregarding the geometry of the input data and the underlying generalized eigenproblem. Considering the linear FEM mass matrix $\mathbf{B}$ and noting that $B(i, j) = \langle \delta_{\mathbf{p}_i}, \delta_{\mathbf{p}_j} \rangle_2$, where $\delta_\mathbf{p}$ is the function that takes value 1 at $\mathbf{p}$ and 0 otherwise, the $\mathbf{B}$-scalar product is the counterpart of the $\mathcal{L}_2(\mathcal{N})$ scalar product on the space of discrete functions on $\mathcal{M}$. The orthogonality of the Laplacian eigenvectors with respect to the $\mathbf{B}$-scalar product is crucial to encode the geometry of the surface underlying $\mathcal{M}$ in the spectral distances and makes its evaluation robust to surface sampling.

Symmetric component of the spectral kernel In some applications, it is useful to work with a symmetric instead of a $\mathbf{B}$-adjoint diffusion kernel matrix. To this end, we introduce the symmetric component $\tilde{\mathbf{K}}_\varphi := \mathbf{X}\varphi(\Lambda)\mathbf{X}^\top$ of the heat kernel matrix, which can be computed through the spectrum-free computation as

$$\tilde{\mathbf{K}}_\varphi\mathbf{f} = \tilde{\mathbf{K}}_\varphi\mathbf{BB}^{-1}\mathbf{f} = \mathbf{K}_\varphi\mathbf{By}, \qquad \mathbf{y} = \mathbf{B}^{-1}\mathbf{f}.$$

Indeed, we evaluate $\mathbf{y}$ as solution to the sparse, symmetric, positive definite linear system $\mathbf{By} = \mathbf{f}$ and we then compute $\mathbf{K}_\varphi(\mathbf{By})$ through the Padé-Chebyshev approximation.

4.2 NATIVE SPECTRAL SPACES

Let $\mathcal{P} := \{\mathbf{p}_i\}_{i=1}^n$ be a set of samples of $\mathcal{N}$ and let us consider the kernel space $\mathcal{K}(\mathcal{N}) := \text{span}\{K_\varphi(\mathbf{p}_i, \cdot)\}_{i=1}^n$ induced by the filter φ and equipped with the $\mathcal{L}_2(\mathcal{N})$ scalar product. In $\mathcal{K}(\mathcal{N})$, the scalar product of two functions $f = \sum_{i=1}^n \alpha_i K_\varphi(\mathbf{p}_i, \cdot), g = \sum_{i=1}^n \beta_i K_\varphi(\mathbf{p}_i, \cdot)$, is expressed as

$$\begin{aligned}
\langle f, g \rangle_2 &= \sum_{i,j=1}^n \alpha_i \beta_j \langle K_\varphi(\mathbf{p}_i, \cdot) K_\varphi(\mathbf{p}_j, \cdot) \rangle_2 \\
&= \sum_{i,j=1}^n \alpha_i \beta_j K_{\varphi^2}(\mathbf{p}_i, \mathbf{p}_j) \\
&= \alpha^\top \tilde{\mathbf{K}}_{\varphi^2}\beta, \quad \tilde{\mathbf{K}}_{\varphi^2} := (K_{\varphi^2}(\mathbf{p}_i, \mathbf{p}_j))_{i,j=1}^n.
\end{aligned} \tag{4.3}$$

In the second equality, we have applied the identity $\langle K_{\varphi_1}(\mathbf{p}_i, \cdot), K_{\varphi_2}(\mathbf{p}_j, \cdot) \rangle_2 = K_{\varphi_1\varphi_2}(\mathbf{p}_i, \mathbf{p}_j)$, which is a consequence of the representation of the spectral kernels and the orthonormality of the Laplacian eigenfunctions. Then, the *native spectral space* $\mathcal{H}$ is defined as the $\| \cdot \|_\mathcal{H}$-closure of functions in $\mathcal{K}(\mathcal{N})$ and the *native scalar product* is

$$\langle f, g \rangle_\mathcal{H} = \sum_{i,j=1}^n \alpha_i \beta_j K_\varphi(\mathbf{p}_i, \mathbf{p}_j) = \alpha^\top \tilde{\mathbf{K}}_\varphi\beta, \tag{4.4}$$

where $\tilde{\mathbf{K}}_\varphi := (K_\varphi(\mathbf{p}_i, \mathbf{p}_j))_{i,j=1}^n$ is the symmetric spectral kernel matrix. Here, the scalar product (4.4) is induced by φ instead of φ^2 (c.f., Eq. (4.3)). Being $\mathbf{f} := (f(\mathbf{p}_i))_{i=1}^n = \tilde{\mathbf{K}}_\varphi \alpha$, we have

$$\|f\|_{\mathcal{H}}^2 = \alpha^\top \tilde{\mathbf{K}}_\varphi \alpha = \mathbf{f}^\top \tilde{\mathbf{K}}_\varphi^{-1} \mathbf{f} = \|\mathbf{S}_\varphi^{-1}\mathbf{f}\|_2^2, \tag{4.5}$$

where $\mathbf{S}_\varphi$ is any matrix such that $\mathbf{S}_\varphi^2 = \tilde{\mathbf{K}}_\varphi$. Considering as $\mathbf{S}_\varphi$ the positive-definite square root $\tilde{\mathbf{K}}_\varphi = \mathbf{V}\mathbf{D}_\varphi\mathbf{V}^\top$, $\mathbf{D}_\varphi := \mathrm{diag}(\varphi(\lambda_i))_{i=1}^n$, we get that $\mathbf{S}_\varphi = \mathbf{V}\mathbf{D}_\varphi^{1/2}\mathbf{V}^\top$.

In the discrete case, the matrix $\mathbf{K}_\varphi = \mathbf{X}\mathbf{D}_\varphi\mathbf{X}^\top\mathbf{B}$ is not symmetric but $\mathbf{B}$-adjoint; indeed, we cannot compute its square root as previously done. To derive a result analogous to Eq. (4.5), the *discrete native scalar product* is $\langle \mathbf{f}, \mathbf{g} \rangle_{\mathcal{H}} = \langle \mathbf{f}, \mathbf{K}_\varphi \mathbf{g} \rangle_{\mathbf{B}} = \langle \mathbf{K}_\varphi \mathbf{f}, \mathbf{g} \rangle_{\mathbf{B}}$ and the corresponding norm is

$$
\begin{aligned}
\|\mathbf{f}\|_{\mathcal{H}}^2 &= \alpha^\top \mathbf{B}\mathbf{K}_\varphi \alpha, \qquad \alpha = \mathbf{K}_\varphi^\dagger \mathbf{f}, \\
&= \mathbf{f}^\top \mathbf{B}\mathbf{D}_\varphi^\dagger \mathbf{X}^\top \mathbf{B}\mathbf{f} \\
&= (\mathbf{X}\mathbf{D}_{\varphi^{1/2}}^\dagger \mathbf{X}^\top \mathbf{B}\mathbf{f})^\top \mathbf{B}(\mathbf{X}\mathbf{D}_{\varphi^{1/2}}^\dagger \mathbf{X}^\top \mathbf{B}\mathbf{f}) \\
&= \|\mathbf{X}\mathbf{D}_{\varphi^{1/2}}^\dagger \mathbf{X}^\top \mathbf{B}\mathbf{f}\|_{\mathbf{B}}^2 \\
&= \|\mathbf{K}_{\varphi^{1/2}}^\dagger \mathbf{f}\|_{\mathbf{B}}^2.
\end{aligned}
$$

Indeed, the discrete native norm of the input scalar function is computed as the $\mathbf{B}$-norm of (i) the vector $\mathbf{K}_{\varphi^{1/2}}^\dagger \mathbf{f}$ through the spectrum-free method or (ii) the solution to the linear system $\mathbf{K}_{\varphi^{1/2}}\alpha = \mathbf{f}$.

Recalling that the spectral kernel is positive definite, the set $\mathcal{B} := \{K_\varphi(\mathbf{p}_i, \cdot) = \mathbf{K}_\varphi \mathbf{e}_i\}_{i=1}^n$ is a basis of the space $\mathcal{F}(\mathcal{M})$ of scalar functions defined on $\mathcal{M}$, where $\mathcal{P} := \{\mathbf{p}_i\}_{i=1}^n$ is the set of points/vertices of $\mathcal{M}$. Identifying the scalar functions f, g with their coefficient vectors α, β (i.e., $\alpha = \mathbf{K}_{1/\varphi}\mathbf{f}$), the $\mathbf{B}$-scalar product in $\mathcal{F}(\mathcal{M})$ is rewritten in terms of the heat kernel as

$$\langle \mathbf{f}, \mathbf{g} \rangle_{\mathbf{B}} = \mathbf{f}^\top \mathbf{B}\mathbf{g} = \alpha^\top \mathbf{K}_\varphi^\top \mathbf{B}\mathbf{K}_\varphi \beta = \alpha^\top \mathbf{B}\mathbf{K}_{\varphi^2} \beta = \langle \alpha, \mathbf{K}_{\varphi^2}\beta \rangle_{\mathbf{B}}, \tag{4.6}$$

where we have applied the identity $\mathbf{K}_\varphi^\top \mathbf{B}\mathbf{K}_\varphi = \mathbf{B}\mathbf{K}_{\varphi^2}$. We notice the analogy between the last equality in (4.6) and (4.3).

4.3   COMPUTATION OF THE SPECTRAL DISTANCES

The analysis of previous work on the computation of the Laplacian spectral kernel and distances will be focused on the following properties:

- *numerical accuracy and stability*; in particular, analysis of the convergence and Gibbs phenomenon. These aspects are important for the computation of the spectral kernels and distances as well as the approximation of "geometry-driven" distances, such as the geodesic and optimal transportation distances;

- *computational cost and storage overhead*, which are crucial for the evaluation of Laplacian spectral kernels and distances on large data sets; and

- *selection of parameters and heuristics*, which generally affects the quality of the final result.

Recalling that the computation of the Laplacian eigenpairs is numerically unstable in case of repeated eigenvalues (Sec. 1.3.2), the filter function should be chosen in such a way that the filtered Laplacian matrix does not have additional (if any) repeated eigenvalues. This condition is generally satisfied by choosing an injective filter. The selection of periodic filters, the expensive cost of the computation of the Laplacian spectrum, and the sensitiveness of multiple Laplacian eigenvalues to surface discretization are the main motivations for the definition of alternative approaches for the evaluation of the spectral distances and kernels. Among them, we discuss the truncated (Sec. 4.3.1) and spectrum-free (Sec. 4.3.2) approximations; then, we introduce a unified approach (Sec. 4.3.3).

4.3.1 TRUNCATED APPROXIMATION

The computational limits for the evaluation of the whole Laplacian spectrum and the decay of the coefficients in Eq. (3.5) are the main reasons behind the approximation of the solution to the spectral distances as a truncated sum, i.e.,

$$\begin{cases} \Phi_k \mathbf{f} = \sum_{i=1}^{k} \varphi(\lambda_i) \langle \mathbf{f}, \mathbf{x}_i \rangle_\mathbf{B} \mathbf{x}_i; \\ d^2(\mathbf{p}_i, \mathbf{p}_j) = \sum_{l=1}^{k} \varphi^2(\lambda_l) |\mathbf{x}_l^\top \mathbf{B} \mathbf{e}_i - \mathbf{x}_l^\top \mathbf{B} \mathbf{e}_j|^2, \end{cases}$$

where k is the number of selected eigenpairs. Even though the first k Laplacian eigenpairs are computed in super-linear time [VL08], the evaluation of the whole Laplacian spectrum is unfeasible for storage and computational cost, which are quadratic in the number of surface samples. Furthermore, the selection of filters that are periodic or do not decrease to zero motivates the need of defining a spectrum-free computation of the corresponding kernels and distances, which cannot be accurately approximated with the contribution of only a subpart of the Laplacian spectrum. The number of selected eigenpairs is heuristically adapted to the decay of the filter function and the approximation accuracy cannot be estimated without computing the whole spectrum.

4.3.2 SPECTRUM-FREE APPROXIMATION

We now introduce the spectrum-free evaluation of the spectral distances, which is based on a polynomial or rational approximation of the filter.

Conditioning of the spectral kernel and computational cost We analyze the conditioning of the spectral kernel. Assuming that φ is an increasing function (i.e., $1/\varphi$ is a low pass filter), the conditioning number of the filtered Laplacian matrix is bounded as

$$\kappa_2(\mathbf{K}_\varphi) = \kappa_2(\varphi(\tilde{\mathbf{L}})) = \frac{\max_{i=1,\dots,n}\{\varphi(\lambda_i)\}}{\min_{i=1,\dots,n}\{\varphi(\lambda_i)\}} = \frac{\|\varphi\|_\infty}{\varphi(0)},$$

and it is ill-conditioned when $\varphi(0)$ is close to zero or φ is unbounded. If φ is bounced and $\varphi(0)$ is not too close to 0, then the filtered Laplacian matrix is well-conditioned. If $\varphi(0)$ is null, then we consider the smallest and not null filtered Laplacian eigenvalue at the denominator of the previous relation.

Analogously to the definitions in Sec. 3.5, the spectral distance is equal to the norm $d(\mathbf{f}, \mathbf{g}) := \|\mathbf{u}\|_{\mathbf{B}}$ of the solution to the linear system $\mathbf{K}_{\varphi}\mathbf{u} = \mathbf{f} - \mathbf{g}$ or to the norm of the vector $\mathbf{u} = \mathbf{K}_{1/\varphi}(\mathbf{f} - \mathbf{g})$. Indeed, this approximation of the spectral distances involves the spectral kernel only and allows us to bypass numerical inaccuracies due to repeated or close Laplacian eigenvalues [GV89, Sec. 7]. Since the eigenvalues of the spectral kernel matrix $\mathbf{K}_{\varphi}$ are $(\varphi(\lambda_i))_{i=1}^{n}$, the distance between the spectral kernel matrices $\mathbf{K}_{\rho}$, $\mathbf{K}_{\varphi}$ reduces to the approximation of the corresponding filters with respect to the ℓ_{∞} norm; in fact,

$$\|\mathbf{K}_{\rho} - \mathbf{K}_{\varphi}\|_2 = \|\mathbf{K}_{\rho-\varphi}\|_2 = \max_{i=1,\dots,n}\{|\rho(\lambda_i) - \varphi(\lambda_i)|\} \leq \|\rho - \varphi\|_{\infty}.$$

The approximation ρ of φ is computed on the interval $[0, \lambda_{\max}(\tilde{\mathbf{L}})]$, where the maximum Laplacian eigenvalue is computed by the Arnoldi method [GV89], or is set equal to the upper bound [LS96, Sor92] $\lambda_{\max}(\tilde{\mathbf{L}}) \leq \min\{\max_i\{\sum_j \tilde{L}(i, j)\}, \max_j\{\sum_i \tilde{L}(i, j)\}\}$.

Arbitrary filter: polynomial approximation   For an arbitrary filter φ, the matrix $\varphi(\mathbf{A})$ is approximated by selecting a new function g such that the matrix $\tilde{\mathbf{A}} = g(\mathbf{A})$ approximates $\mathbf{A}$ and can be easily calculated. One of the main approaches for the approximation of a matrix function is through the truncated Taylor approximation [GV89]. More precisely, given the power series representation $\varphi(s) = \sum_{n=0}^{+\infty} \alpha_n s^n$ defined on an open disk containing the spectrum of $\mathbf{A}$, we have that $\varphi(\mathbf{A}) = \sum_{n=0}^{+\infty} \alpha_n \mathbf{A}^n$. In this case, it is enough to consider the contribution of the first k terms in the sum and to compute the powers $(\mathbf{A}^i)_{i=1}^{k}$, through a binary powering [VL79].

Let $[0, \lambda]$ be an interval that contains the spectrum of $\tilde{\mathbf{L}}$, where λ is the maximum eigenvalue, which is computed by the Arnoldi method [GV89], or is set equal to the upper bound [LS96, Sor92] $\lambda_n \leq \min\{\max_i\{\sum_j \tilde{L}(i, j)\}, \max_j\{\sum_i \tilde{L}(i, j)\}\}$. Applying the Taylor approximation $\varphi(s) \approx p_r(s) := \sum_{n=0}^{r} \alpha_n s^n$ to the Laplacian matrix in $[0, \lambda]$, $\mathbf{K}\mathbf{e}_i$ is evaluated as (Algorithm 4.2)

$$\mathbf{K}\mathbf{e}_i \approx \sum_{n=0}^{r} \alpha_n (\mathbf{B}^{-1}\mathbf{L})^n \mathbf{e}_i = \alpha_0 \mathbf{e}_i + \sum_{n=1}^{r} \alpha_n \mathbf{g}_n, \qquad (4.7)$$

where $\mathbf{g}_n$ satisfies the linear system $\mathbf{B}\mathbf{g}_{n+1} = \mathbf{L}\mathbf{g}_n$, $\mathbf{B}\mathbf{g}_1 = \mathbf{L}\mathbf{e}_i$.

From the upper bound [Pat14]

$$\left\|\varphi(\tilde{\mathbf{L}}) - \sum_{n=0}^{r} \alpha_n \tilde{\mathbf{L}}^n\right\|_2 \leq \frac{n}{(r+1)!}\left[\frac{\lambda_{\max}(\mathbf{L})}{\lambda_{\min}(\mathbf{B})}\right]^{r+1}\|\varphi^{(r+1)}(\tilde{\mathbf{L}})\|_2,$$

it follows that the approximation accuracy is mainly controlled by the degree of the Taylor approximation and the variation of the ratio between the maximum eigenvalue of $\mathbf{L}$ and the minimum eigenvalue of $\mathbf{B}$. If necessary, a higher approximation accuracy is achieved by slightly

---

Algorithm 4.2 Computation of the spectral distances.

Require: A surface or volume $\mathcal{M}$, a filter function $\varphi : \mathbb{R} \to \mathbb{R}$.
Ensure: The spectral distance $d(\mathbf{p}_i, \mathbf{p}_j)$ in Eq. (4.1), $\mathbf{p}_i, \mathbf{p}_j \in \mathcal{M}$.
1: Compute $(\mathbf{L}, \mathbf{B})$, which define the Laplacian $\tilde{\mathbf{L}} := \mathbf{B}^{-1}\mathbf{L}$.
2: Define the vector $\mathbf{f} = \mathbf{e}_i - \mathbf{e}_j$.

3: *CASE I—Arbitrary filter: polynomial approximation*
4: Compute the polynomial approx. $p_r(s) = \sum_{i=0}^{r} \alpha_i s^i$ of φ.
5: Compute $\mathbf{g}_1$: $\mathbf{B}\mathbf{g}_1 = \mathbf{L}\mathbf{f}$.
6: **for** $i = 1, \ldots, r-1$ **do**
7:     Compute $\mathbf{g}_{i+1}$: $\mathbf{B}\mathbf{g}_{i+1} = \mathbf{L}\mathbf{g}_i$
8: **end for**
9: Compute $\mathbf{u} = \mathbf{K}\mathbf{f} \approx p_r(\tilde{\mathbf{L}}) = \alpha_0 \mathbf{f} + \sum_{i=1}^{r} \alpha_i \mathbf{g}_i$ (c.f., Eq. (4.7)).
10: Compute the distance $d(\mathbf{p}_i, \mathbf{p}_j) = \|\mathbf{u}\|_{\mathbf{B}}$.

11: *CASE II - Arbitrary filter: Padé-Chebyshev approximation*
12: Compute the P.C. approx. $p_r(s) = \sum_{i=1}^{r} \alpha_i (1 + \beta_i s)^{-1}$ of φ.
13: **for** $i = 1, \ldots, r$ **do**
14:     Compute $\mathbf{g}_i$: $(\mathbf{B} + \beta_i \mathbf{L})\mathbf{g}_i = \mathbf{B}\mathbf{f}$ (c.f., Eq. (4.8))
15: **end for**
16: Compute $\mathbf{u} = \mathbf{K}\mathbf{f} \approx p_r(\tilde{\mathbf{L}})\mathbf{f} = \sum_{i=1}^{r} \alpha_i \mathbf{g}_i$.
17: Compute the distance $d(\mathbf{p}_i, \mathbf{p}_j) = \|\mathbf{u}\|_{\mathbf{B}}$.

---

increasing the degree r. Finally, this computation of both the spectral kernel and distance is independent of the discretization of the input surface as a polygonal mesh or a point cloud. In case of a complex kernel, it is enough to apply the previous discussion to its real and imagery parts; e.g., for the wave kernel we consider the series $\sin(\tilde{\mathbf{L}}) = \sum_{n=0}^{+\infty}(-1)^n \tilde{\mathbf{L}}^{2n+1}/(2n+1)!$ and $\cos(\tilde{\mathbf{L}}) = \sum_{n=0}^{+\infty}(-1)^n \tilde{\mathbf{L}}^{2n}/(2n)!$.

Arbitrary filter: Padé-Chebyshev approximation For an arbitrary filter, we consider the rational Padé-Chebyshev approximation $p_r(s) = \frac{a_r(s)}{b_r(s)}$ of φ [GV89, Ch. 11] with respect to the $\mathcal{L}_\infty$ norm. Here, $a_r(\cdot)$ and $b_r(\cdot)$ are polynomials of degree equal to or lower than r. Let $p_r(s) = \sum_{i=1}^{r} \alpha_i (1 + \beta_i s)^{-1}$ be the partial form of the Padé-Chebyshev approximation, where $(\alpha_i)_{i=1}^{r}$ are the weights and $(\beta_i)_{i=1}^{r}$ are the nodes of the r-point Gauss-Legendre quadrature rule [GV89, Ch. 11]. The weights and nodes are precomputed for any degree of the rational polynomial [CRV84]. Applying this approximation to the spectral kernel, we get that

$$\mathbf{u} = \mathbf{K}\mathbf{f} \approx p_r(\tilde{\mathbf{L}})\mathbf{f} = \sum_{i=1}^{r} \alpha_i \left(\mathbf{I} + \beta_i \tilde{\mathbf{L}}\right)^{-1} \mathbf{f} = \sum_{i=1}^{r} \alpha_i \mathbf{g}_i,$$

where $\mathbf{g}_i$ solves the symmetric and sparse linear system

$$(\mathbf{B} + \beta_i \mathbf{L})\mathbf{g}_i = \mathbf{B}\mathbf{f}, \quad i = 1, \ldots, r. \tag{4.8}$$

The Padé-Chebyshev approximation generally provides an accuracy higher than the polynomial approximation, as a matter of its uniform convergence to the filter.

Properties According to [MVL03], the approximation of the matrix $\varphi(\tilde{\mathbf{L}})$ might be numerically unstable if $\|\tilde{\mathbf{L}}\|_2$ is large. From the bound $\|\mathbf{B}^{-1}\mathbf{L}\|_2 \leq \lambda_{\min}^{-1}(\mathbf{B})\lambda_{\max}(\mathbf{L})$, a well-conditioned mass matrix $\mathbf{B}$ guarantees that $\|\mathbf{B}^{-1}\mathbf{L}\|_2$ is bounded. Recalling that $\mathbf{X}^\top(\mathbf{B} + \beta_i\mathbf{L})\mathbf{X} = (\mathbf{I} + \beta_i\Lambda)$, $\{1 + \beta_i\lambda_j\}_{j=1}^n$ are the eigenvalues of $(\mathbf{B} + \beta_i\mathbf{L})$ and its conditioning number is bounded by the constant $(1 + \beta_{\max}\lambda_n)$, $\beta_{\max} := \max_{i=1,\dots,n}|\beta_i|$. Indeed, the coefficient matrices in Eq. (4.7) are well-conditioned and specialized pre-conditioners [KFS13] can be applied to further attenuate numerical instabilities.

Approximating an arbitrary filter function with a rational or a polynomial function of degree r, the evaluation of the corresponding spectral distance between two points is reduced to solve r sparse, symmetric, linear systems (c.f., Eq. (4.7)), whose coefficient matrices have the same structure and sparsity of the connectivity matrix of the input triangle mesh or of the k-nearest neighbor graph for a point set. Applying iterative solvers, such as the Jacobi, Gauss-Seidel, minimum residual methods [GV89], and without extracting the Laplacian spectrum, the computational cost is $\mathcal{O}(r\tau(n))$, where $\tau(n)$ is the cost for the solution of a sparse linear system, which varies from $\mathcal{O}(n)$ to $\mathcal{O}(n^2)$, according to the sparsity of the coefficient matrix, and it is $\mathcal{O}(n \log n)$ in the average case.

The spectrum-free computation of the one-to-all distances $\{d(\mathbf{p}_i, \mathbf{p}_j)\}_{j=1}^n$ takes $\mathcal{O}(rn\tau(n))$ time; in fact, we solve the sparse linear system (4.7) with n different right-hand vectors $(\mathbf{e}_i - \mathbf{e}_j)$, $j = 1, \dots, n$. Computing a fixed number k of eigenpairs in $\mathcal{O}(kn)$ time, the truncated spectral approximation of the one-to-all distance is evaluated in constant time for any filter. Indeed, the spectrum-free approach is competitive with respect to the truncated spectral approximation with $k(n) \geq r\tau(n)$ Laplacian eigenpairs. In the average case, $\tau(n) \approx n \log n$ and $k(n) \geq k_n$, $k_n = r \log n$. For instance, for a surface with $n = 10^4, 10^5, 10^6$ points and a degree $r = 5$, the number of eigenpairs is $k_n = 46, 58, 69$; in particular, this growth of k_n with respect to n is slow, as a matter of the logarithm in k_n.

4.3.3 A UNIFIED SPECTRUM-FREE COMPUTATION

We now propose a novel and efficient computation of the spectral distances induced by an arbitrary kernel, which integrates the methods previously discussed. This approximation is based on the recursive solution of a set of sparse, symmetric linear systems and is induced by the representation of the rational polynomial in terms of a the basis $\mathcal{B} := \{s^i\}_{i=0}^l \cup \{s^{-i}\}_{i=1}^r$. It also generalizes the one proposed in [Pat16a], which depends on the polynomial or rational approximation of the input filter.

Recalling that the dimension of the space $\mathcal{R}_r^l$ of rational polynomial of degree (r, l) is $l + r + 1$, let us express the best rational approximation c_{rl} of φ in terms of these basis functions. To this end, we can either apply algebraic rules or impose interpolating constraints at $(l + r + 1)$

points in order to identify the new coefficients through the identity

$$\sum_{i=0}^{l} a_i s^i + \sum_{i=1}^{r} b_i s^{-i} = c_{rl}(s).$$

The basis $\mathcal{B}$ used to represent the rational approximation of the input filter simplifies the computation of the spectral kernel; in fact, we need to compute only the following vectors $\tilde{\mathbf{L}}^i \mathbf{f}$ and $(\tilde{\mathbf{L}}^i)^\dagger \mathbf{f}$. Applying the spectral representation of the Laplacian matrix

$$\tilde{\mathbf{L}} = \mathbf{X}\Lambda\mathbf{X}^\top\mathbf{B}, \quad \Lambda = \mathrm{diag}(\lambda_i)_{i=1}^n, \quad 0 = \lambda_1 < \lambda_i, \quad i = 2,\ldots,n,$$

we can represent its powers and the corresponding pseudo-inverse matrices as

$$\tilde{\mathbf{L}}^i = \mathbf{X}\Lambda^i\mathbf{X}^\top\mathbf{B}, \quad (\tilde{\mathbf{L}}^i)^\dagger = \mathbf{X}(\Lambda^i)^\dagger\mathbf{X}^\top\mathbf{B}, \quad \Lambda^\dagger = \mathrm{diag}(0, \lambda_i^{-1})_{i=2}^n.$$

To compute $\mathbf{g}_1 = \tilde{\mathbf{L}}^\dagger \mathbf{f}$, let us multiply both sides of this equation by $\tilde{\mathbf{L}}$ and notice that

$$\begin{aligned}
\tilde{\mathbf{L}}\mathbf{g}_1 &= \tilde{\mathbf{L}}\tilde{\mathbf{L}}^\dagger \mathbf{f} \\
&= \mathbf{X}\Lambda\Lambda^\dagger\mathbf{X}^\top\mathbf{B}\mathbf{f}, \quad \mathbf{X}^\top\mathbf{B}\mathbf{X} = \mathbf{I}, \\
&= \mathbf{X}(\mathbf{I} - \mathbf{e}_1\mathbf{e}_1^\top)\mathbf{X}^\top\mathbf{B}\mathbf{f}, \quad \mathbf{e}_1 := [1, 0, \ldots, 0]^\top, \\
&= \mathbf{f} - (\mathbf{1}^\top\mathbf{B}\mathbf{f})\mathbf{1}, \quad \mathbf{1} := [1, 1, \ldots, 1]^\top,
\end{aligned}$$

i.e., $\mathbf{g}_1$ is the least-squares solution to the sparse and symmetric linear system $\mathbf{L}\mathbf{g}_1 = \mathbf{B}\mathbf{f}_1$, $\mathbf{f}_1 := \mathbf{f} - (\mathbf{1}^\top\mathbf{B}\mathbf{f})\mathbf{1}$, where $\mathbf{f}_1$ is achieved by subtracting to $\mathbf{f}$ its mean value $\langle \mathbf{f}, \mathbf{1} \rangle_\mathbf{B}$. For the general case, we apply the recursive relation

$$\mathbf{g}_i := (\tilde{\mathbf{L}}^\dagger)^i \mathbf{f} = \tilde{\mathbf{L}}^\dagger(\tilde{\mathbf{L}}^\dagger)^{i-1}\mathbf{f} = \tilde{\mathbf{L}}^\dagger\mathbf{g}_{i-1}, \quad i \geq 2,$$

which reduces to the previous case (Fig. 4.1). In a similar way, $\mathbf{g}_i = \tilde{\mathbf{L}}^i \mathbf{f}$ is calculated by recursively solving the sparse, symmetric, and positive-definite linear systems $\mathbf{B}\mathbf{g}_{i+1} = \mathbf{L}\mathbf{g}_i$, $i = 1, \ldots, r - 1$, with $\mathbf{g}_0 := \mathbf{f}$.

4.4   DISCUSSION

Recalling that an arbitrary filter $1/\varphi$ can be approximated in $\mathcal{R}_r^l$ with an accuracy of order $\mathcal{O}(s^{l+r+1})$ with respect to the ℓ_∞-norm (Sec. 4.3.3), we can use the space $\mathcal{R}_r^l$ to define any filter and the corresponding spectral distances (Fig. 4.1). In this way, we reduce the degree of freedom in the definition of the filter to the selection of $(l + r + 1)$ coefficients and without loosing the richness of the resulting spectral distances. For the selection of the coefficients of the rational filter and the filter frequencies, we can apply the rules proposed in [KR05] for Laplacian spectral smoothing. In particular, the filter frequencies are derived from the dimension of a bounding box placed around the chosen feature $\mathcal{F}$ and whose axis are aligned with the eigenvectors of the covariance matrix of $\mathcal{F}$.

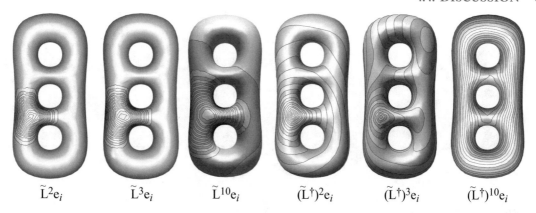

$\widetilde{L}^2 e_i$ $\qquad$ $\widetilde{L}^3 e_i$ $\qquad$ $\widetilde{L}^{10} e_i$ $\qquad$ $(\widetilde{L}^\dagger)^2 e_i$ $\qquad$ $(\widetilde{L}^\dagger)^3 e_i$ $\qquad$ $(\widetilde{L}^\dagger)^{10} e_i$

Figure 4.1: Level sets of the basis functions at $\mathbf{p}_i$ associated with the rational polynomial approximation.

For an arbitrary 3D shape, we cannot compute the ground-truth diffusion distances through the analytic representation of its Laplacian eigenfunctions, as done for the sphere and cylinder (Fig. 4.2, Fig. 4.3). Recalling the upper bound to the accuracy of the Padé-Chebyshev approximation of the diffusion distances with respect to the selected degree of the rational polynomial, we can analyze the discrepancy of this approximation with respect to the truncated spectral approximation. For this example, we report the ℓ_∞ error (y-axis) between the Padé-Chebyshev approximation of the heat kernel at different scales and the truncated spectral approximation with a different number (x-axis) of Laplacian eigenpairs. At large scales (e.g., $t = 10^{-1}, 1$), the truncated spectral approximation with a low number of eigenpairs (e.g., $k \approx 50$) provides an approximation of the solution to the heat equation that is close to the one computed by the Padé-Chebyshev approximation, with respect to the ℓ_∞-norm ($\epsilon_\infty = 10^{-4}$).

The level sets of these two approximations also show their different behavior at small scales, while at larger scales it becomes analogous in terms of both the shape and distribution of the level sets. Finally, at small scales (e.g., $t = 10^{-2}, 10^{-2}$) the truncated spectral approximation is generally affected by small undulations far from the seed point and that do not disappear while increasing the number of Laplacian eigenpairs. Increasing the number of Laplacian eigenpairs makes these undulation more evident, as a matter of the small vibrations of the Laplacian eigenvectors associated with larger Laplacian eigenvalues, which are included in the truncated approximation. In this case, a larger number of eigenpairs is necessary to obtain a truncated spectral approximation close to the one computed with the Padé-Chebyshev rational polynomial. Indeed, this situation is analogous to the results achieved on the sphere and the cylinder.

As shown in Fig. 4.4, the approximation of the diffusion, and more generally, spectral distances are affected by small undulations (especially at small scales), the use of heuristics for the selection of the number of Laplacian eigenpairs with respect to the target approximation

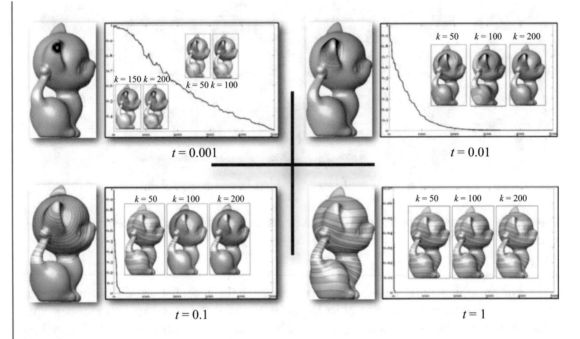

Figure 4.2: ℓ_∞ error (y-axis) between the Padé-Chebyshev approximation of the heat kernel at different scales and the truncated spectral approximation with a different number (x-axis) of Laplacian eigenpairs.

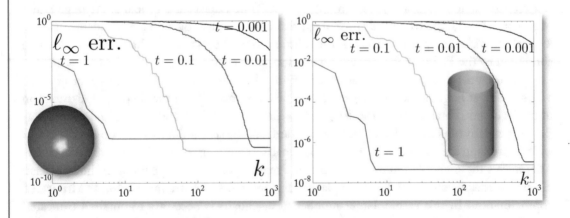

Figure 4.3: ℓ_∞ error (y-axis) for the diffusion distance approximated with k (x-axis) Laplacian eigenpairs. For the Padé-Chebyshev method ($r = 5$) and all the scales, the ℓ_∞ error with respect to the ground-truth is lower than 8.9×10^{-6}.

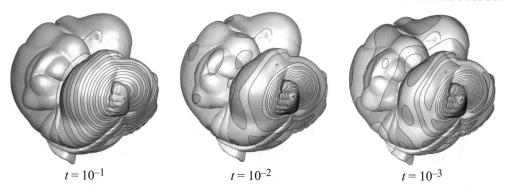

$t = 10^{-1}$ $t = 10^{-2}$ $t = 10^{-3}$

Figure 4.4: Numerical instabilities of the spectral truncated approximation of the heat kernel at small scale. (See also Fig. 4.5).

accuracy and the scale of features of the input shape, the overall computational cost and storage overhead for the computation of a subpart of the Laplacian spectrum. The comparison of these results with the ones induced by the spectrum-free computation (Fig. 4.5) show the improvement of the proposed approach in terms of smoothness, regularity, and accuracy of the computed distances.

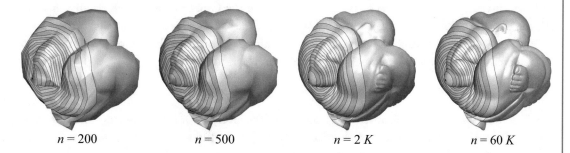

$n = 200$ $n = 500$ $n = 2\,K$ $n = 60\,K$

Figure 4.5: Robustness of the Padé-Chebyshev approximation of the heat kernel with respect to a different resolution (n vertices) of the input shape.

The truncated spectral approximation of the diffusion kernel is generally affected by the Gibbs phenomenon, i.e., small negative distance values. This phenomenon is more evident at small cases, which induce diffusion kernel that decrease fast to zero and that are largely affected by small negative values. In fact, at small scales the kernel values decrease fast to zero and the negative values are no more compensated by the Laplacian eigenvectors related to smaller eigenvalues, as they are not included in the approximation (Fig. 4.6e,f). For the Padé-Chebyshev approximation (Fig. 4.6(a-d)), the distance values are positive at all the scales; in fact, we approximate the filter function without selecting a sub-part of the Laplacian spectrum.

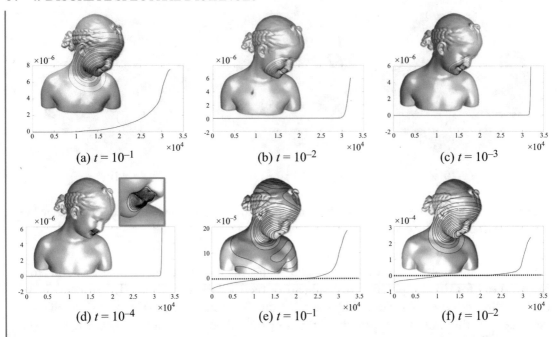

Figure 4.6: (a–d) Robustness of the Padé-Chebyshev approximation of the diffusion kernel and (e,f) sensitiveness of truncated spectral approximation to the Gibbs phenomenon. At all scales (a–d), the distance values (red curve) computed with the Padé-Chebyshev approximation are positive; at large scales (e,f), the truncated spectral approximation is affected by the Gibbs phenomenon, as represented by the part of the plot below the zero line (black curve).

In our experiments, the analogous behavior of the level sets of the heat kernel and diffusion distance confirm the robustness of the Padé-Chebyshev of the approximation with respect to surface resolution (Fig. 4.5), the truncated spectral approximation (Fig. 4.6), partial sampling (Fig. 4.7), geometric (Fig. 4.8) and topological (Fig. 4.9) noise, deformations (Fig. 4.10), and surface discretization as triangle meshes and point sets (Fig. 2.11). A higher resolution of $\mathcal{M}$ improves the quality of the level sets, which are always uniformly distributed and an increase of the noise magnitude does not affect the shape and distribution of the level sets.

Figure 4.3 reports the ℓ_∞ discrepancy (y-axis) between the diffusion distance on the sphere/cylinder and its approximation computed with the Padé-Chebyshev method and the truncated spectral approximation. In this case, the analytical expression of the Laplacian eigenfunctions on the sphere and cylinder has been used to compute the ground-truth distances [Pat16a]. For small scales (e.g., $t = 10^{-2}$, 10^{-3}), the approximation error remains higher than 10^{-2}, with $k \leq 280$ eigenpairs; in fact, local shape features encoded by the heat kernel are recovered for a small t using the eigenvectors associated with high frequencies, thus requiring

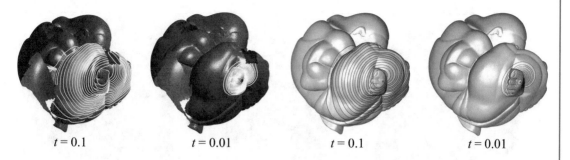

$t = 0.1$ $t = 0.01$ $t = 0.1$ $t = 0.01$

Figure 4.7: Stability of the Padé-Chebyshev computation of the diffusion kernel with respect to partially sampled surfaces.

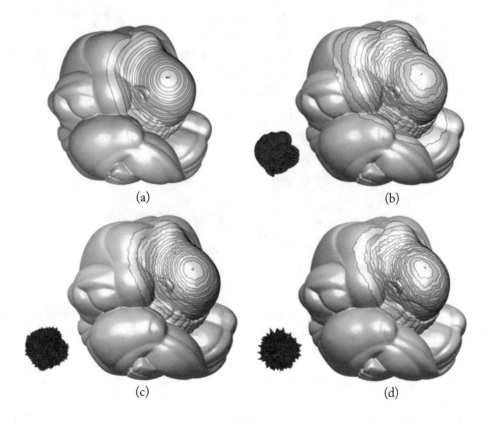

(a)   (b)

(c)   (d)

Figure 4.8: Robustness of the computation of the linear FEM heat kernel from a seed point placed on the leg. Level sets of the diffusion distances on (b–d) the noisy shapes (bottom part, red surfaces) have been plotted on (a) the initial shape.

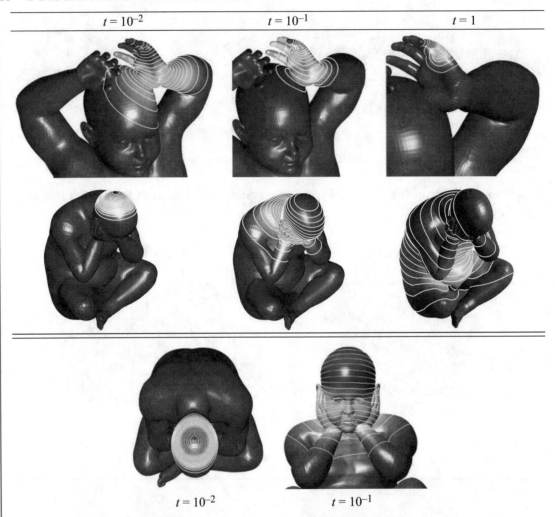

| $t = 10^{-2}$ | $t = 10^{-1}$ | $t = 1$ |

$t = 10^{-2}$ $t = 10^{-1}$

Figure 4.9: Robustness of the Padé-Chebyshev computation of the diffusion kernel at different scales with respect to (first, second row) topological noise (third row). Zoom-in.

the computation of a large part of the Laplacian spectrum. For large scales (e.g., $t = 1$, 10^{-1}), increasing k strongly reduces the approximation error until it becomes almost constant and close to zero. In this case, the behavior of the heat kernel is mainly influenced by the Laplacian eigen-vectors related to the smaller eigenvalues. Indeed, the truncated spectral representation generally requires a high number of eigenpairs and does not achieve the approximation accuracy of our approach, which remains lower than 8.9×10^{-6} for all the scales. According to [VBCG10], there are no theoretical guarantees on the approximation accuracy of the heat kernel provided

Figure 4.10: Robustness of the computation of the linear FEM heat kernel with respect to shape deformations.

by multi-resolution prolongation operators. Furthermore, a low-resolution sampling of the input surface might affect the resulting accuracy.

Results in Fig. 4.11 and Fig. 4.12 confirm that the diffusion distances at small scales generally require a number of eigenpairs that is much higher than the estimated value k_n. All tests have been performed on a 2.7 GHz Intel Core i7 Processor, with 8 GB memory. This case makes our computation of the one-to-all distance competitive with respect to its truncated approximation and useful to evaluate the distances for slowly increasing (e.g., diffusion distances at small scales) or periodic filters or among seed points, as if happens for the evaluation of shape descriptors [OFCD02] and bags-of-features [BB11a, BBOG11]. Here, the number of seeds is much lower than the number of samples and the higher accuracy of our computation improves the discrimination capabilities of descriptors based on spectral distances.

Fixing the number of Laplacian eigenpairs makes the truncated spectral approximation of the one-to-all distances faster than ours but generally provides a lower approximation accuracy. Slowly increasing filters and small scales for the diffusion distances also require the computation of a large number of Laplacian eigenpairs, thus reducing the gap between the computational cost of the proposed approximation of the one-to-all distances and previous work. An analogous discussion applies to prolongation operators, which compute the truncated spectral approximation on a lower resolution of the input shape. Furthermore, previous work has not addressed methods for the selection of the proper number of eigenpairs with respect to the target approximation accuracy, which cannot be estimated without computing the whole Laplacian spectrum. Finally, Fig. 2.11 and Fig. 4.13 show the robustness of the spectrum-free computation with respect to a

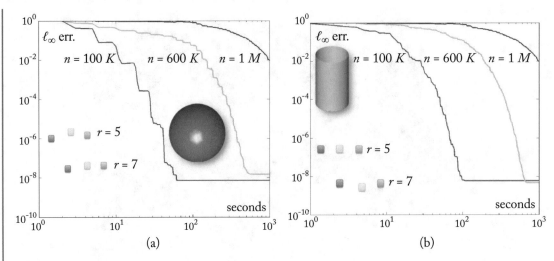

(a)   (b)

Figure 4.11: Trade-off between accuracy (y-axis) and time (x-axis) for the Padé-Chebyshev ($r = 5, 7$) and truncated approximations ($k = 50$ eigenpairs) on the (a) sphere and (b) cylinder.

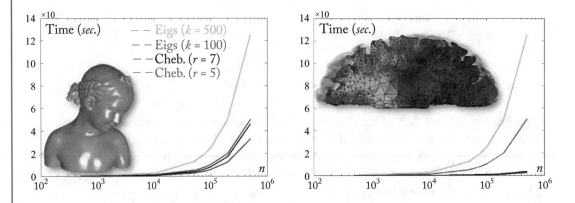

Figure 4.12: Timings (in seconds) for the evaluation of the heat kernel on a domain with n points, approximated with $k = 100, 500$ eigenpairs *(Eigs)*, and the Padé-Chebyshev approximation ($r = 7$).

different shape discretization, non-manifold and bordered surfaces, topological noise. At large scales only (e.g., $t = 1$), the shape of the level sets of the heat kernel changes in a neighbor of the topological cut.

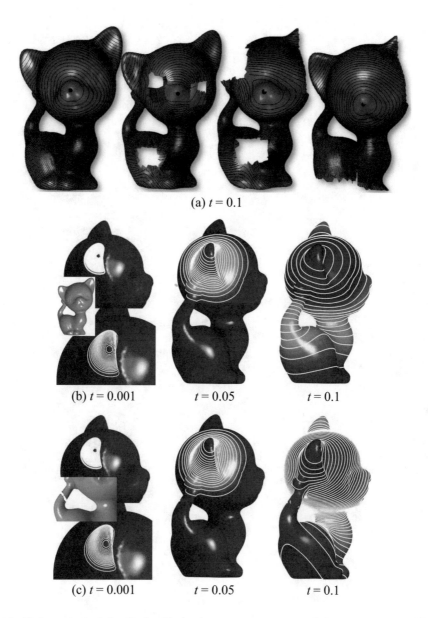

(a) $t = 0.1$

(b) $t = 0.001$ $t = 0.05$ $t = 0.1$

(c) $t = 0.001$ $t = 0.05$ $t = 0.1$

Figure 4.13: Robustness of the Padé-Chebyshev approximation of the linear FEM (a) diffusion distance on partially sampled surfaces and (b,c) heat kernel on smooth and topologically noisy surfaces (cut on the kitten tail), respectively.

CHAPTER 5

Applications

In several applications (e.g., engineering, geographical applications, bio-informatics, and scientific visualization), scalar functions defined on a surface or a volume are used to model a relevant information about the input data or an underlying phenomenon. Controlling the distribution of the critical points (i.e., maxima, minima, saddles) of a scalar function during its design is also crucial for geometry processing and shape analysis. For instance, properly designing a scalar function with a prescribed set of critical points provides a flexible control over the number, shape, and size of the resulting quadrangular patches of remeshed surfaces [DBG+06, HZM+08, NGH04], the number of nodes of the Reeb graph [PSBM07, PSF09] and of the Morse-Smale complex. However, harmonic functions allow the user to select only the number and position of its maxima and minima, with no control on the corresponding saddles. In a similar way, Laplacian eigenfunctions associated with small eigenvalues and diffusion maps at small scales are characterized by a generally low number of critical points, whose spatial location cannot be defined *a-priori*.

Our first goal (Sec. 5.1) is to design a smooth function whose maxima, minima, and saddles are selected by the user or imported from a template function. As main examples of template functions, we mention the Laplacian eigenfunctions [RWSN09] and the diffusion maps [BB11b, Pat17a], which are intrinsically defined by the input surface and their critical points represent relevant shape features, such as protrusions and high-curvature regions. The function with designed critical points is computed by combining interpolating or least-squares constraints with the spectral properties of the Laplacian matrix. As novel contribution [Pat17b] with respect to previous work, we allow the user to select the saddles of the designed function, and not only its extrema.

Then, we will address the Laplacian smoothing of noisy scalar functions, without and with constraints on the preservation of their critical points (Sec. 5.2).

5.1 DESIGN OF SCALAR FUNCTIONS WITH CONSTRAINED CRITICAL POINTS

We aim at designing a smooth function whose critical points are selected by the user or imported from a template function, such as the Laplacian and the diffusion functions. The function with designed critical points is computed by applying interpolating and least-squares constraints and solving a linear system. In all the paper examples, the values of the input scalar function belong to the interval $[0, 1]$. In a similar way, the coordinates of the vertices of the input surface are normalized in such a way that the surface belongs to the unitary bounding box.

Critical point classification   Given a $\mathcal{C}^1$ function $f : \mathcal{M} \to \mathbb{R}$ defined on a smooth 2-manifold surface $\mathcal{M}$, the *critical points* of f are defined as those points $\mathbf{p} \in \mathcal{M}$ such that $\nabla f(\mathbf{p}) = \mathbf{0}$ and they correspond to the maxima, minima, and saddles of f. For polyhedral surfaces, the method described in [Ban67a] classifies a vertex according to the values of f on its neighborhod. To this end, we assume that f is general (i.e., $f(\mathbf{p}_i) \neq f(\mathbf{p}_j)$, $i \neq j$). This assumption guarantees that the level sets of f related to regular iso-values are not degenerate and the Euler formula applies (c.f., Eq. (5.2)). The critical points of a scalar function defined on a triangle mesh can be located and classified by analysing for each vertex $\mathbf{p}_i$ the distribution of the f-values on the neighborhoods of $\mathbf{p}_i$ [Ban67b]. More precisely, let $\mathcal{N}(i) := \{j : (i, j) \text{ edge}\}$ be the 1-*star* of i, i.e., the set of vertices incident to i. The *link* $Lk(i)$ of i is achieved by reordering the indices of $\mathcal{N}(i)$ in an anti-clockwise manner; assuming for simplicity that the indices of $Lk(i)$ are $\{1, \ldots, k\}$, $Lk(i)$ is defined as

$$Lk(i) := \{j \in \mathcal{N}(i) : (j, j+1)_{j=1}^{k-1} \text{ edge of } \mathcal{M}\}.$$

Then, the *upper link* is defined as

$$Lk^+(i) := \{j \in Lk(i) : f(\mathbf{p}_j) > f(\mathbf{p}_i)\},$$

and the *mixed link* as

$$Lk^{\pm}(i) := \{j \in Lk(i) : f(\mathbf{p}_{j+1}) > f(\mathbf{p}_i) > f(\mathbf{p}_j)\}, \tag{5.1}$$

where j is intended as mod $(k + 1)$. The *lower link* $Lk^-(i)$, is defined by replacing the inequality ">" with "<" in the definition of the upper link. If $Lk^+(i) = \emptyset$ or $Lk^-(i) = \emptyset$, then $\mathbf{p}_i$ is a *maximum* or a *minimum*, respectively. If the cardinality of the set $Lk^{\pm}(i)$ is $2 + 2m_i$, $m_i \geq 1$, then $\mathbf{p}_i$ is classified as a *saddle* of *multiplicity* m_i. Once the vertex-vertex relation has been extracted, the classification procedure requires $\mathcal{O}(n)$-time.

Under the assumption that $\mathcal{M}$ is a closed surface, the *Euler formula*

$$\chi(\mathcal{M}) = m - s + M, \quad g = \frac{1}{2}(2 - \chi(\mathcal{M})), \tag{5.2}$$

gives the link between the critical points of $(\mathcal{M}, f)$ and the Euler characteristic $\chi(\mathcal{M})$ of $\mathcal{M}$ [Ban67b, Mil63]. Here, m and M is the number of minima and maxima; the $s := \sum_{\mathbf{p}_i \text{ saddle}} m_i$ saddle points of f are counted with their multiplicity m_i.

Selection of the critical points and values   Given a triangle mesh $\mathcal{M}$ with $\mathcal{P} := \{\mathbf{p}_i\}_{i=1}^n$ set of vertices, we want to design a piecewise linear function $f : \mathcal{M} \to \mathbb{R}$ by selecting the position and values of its critical points. To define a function f with a maximum at a vertex $\mathbf{p}_i$

(Fig. 5.1a), it is enough to define $f(\mathbf{p}_i)$ greater than the f-values at the points of its 1-star $\mathcal{N}(i) := \{j : (i, j) \text{ edge}\}$. Recalling that the cardinality of the mixed link of a saddle of multiplicity m is $2 + 2m$, a vertex $\mathbf{p}_i$ with a neighboring vertices locates a saddle of multiplicity lower than or equal to $\lfloor (a - 2)/2 \rfloor$, where $\lfloor \cdot \rfloor$ is the floor symbol. If this condition is not satisfied, then we either reduce the multiplicity of the saddle according to the previous relation, or we locally update the mesh connectivity at $\mathcal{N}(i)$ (e.g., by splitting each triangle incident to i into two sub-triangles). Indeed, the f-values at $\mathbf{p}_i$ and at its neighboring points are chosen in such a way that the conditions in Eq. (5.1) are satisfied.

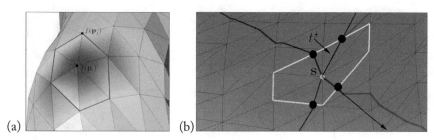

(a)   (b)

Figure 5.1: (a) 1-star of a vertex used for the design and classification of the critical points; maximum achieved by choosing the f-values such that $f(\mathbf{p}_j) > f(\mathbf{p}_i)$, $j \in \mathcal{N}(i)$. (b) Neighbor and level-set (blue, red curves) at a saddle s and points (black dot) that define the mixed link (c.f., Eq. (5.1)). Here, $t^\star$ is a triangle of the neighbor of s intersected by the corresponding level-set.

Let $\mathcal{C}$ be the set of critical points (i.e., maxima, minima, saddles) and let us consider the set $\tilde{\mathcal{I}} := \{j \in \mathcal{N}(i), \ i \in \mathcal{C}\}$ of the vertices of the 1-stars of the critical points. Then, let $\mathcal{I} := \mathcal{C} \cup \tilde{\mathcal{I}}$ be the set of designed critical points and of the corresponding 1-stars' vertices, without repetitions; we also assume that k is the cardinality of the set $\mathcal{I}$. To define the function $f : \mathcal{P} \to \mathbb{R}$ at all the points of $\mathcal{P}$ starting from the designed critical points and the values $\{f(\mathbf{p}_i)\}_{i \in \mathcal{I}}$, we reformulate the problem as the computation of a function that satisfies the conditions $f(\mathbf{p}_i) = f_i$, $i \in \mathcal{I}$, in an exact or least-squares way. By definition, the set $\tilde{\mathcal{C}}$ of critical points of f contains $\mathcal{C}$.

Design of scalar functions with constrained critical points   According to the previous discussion, the smooth scalar function $f : \mathcal{M} \to \mathbb{R}$, with a set $\mathcal{C}$ of critical points, is defined as the solution of the constrained minimization problem (Fig. 5.2)

$$\min_{\mathbf{f} \in \mathbb{R}^n}\{\|\mathbf{Lf}\|_2\}, \qquad f(\mathbf{p}_i) := f_i, \quad i \in \mathcal{I}, \tag{5.3}$$

where $\mathbf{L}$ is the Laplacian matrix with Voronoi-cotangent weights [DMSB99, PP93]. We briefly recall that the Laplacian matrix discretizes the Laplace-Beltrami operator and $\|\mathbf{Lf}\|_2$ represents the Dirichlet energy of f. Since this last term is null or close to zero for constant or smooth functions, its minimization is aimed at controlling the oscillations of the solution to (5.3). To compute the solution to Eq. (5.3), we consider the complement $\mathcal{I}^C$ of $\mathcal{I}$ and for $i \in \mathcal{I}^C$ we have

Figure 5.2: (a,c,e,g) Critical points and (b,d,f,h) of the designed scalar function. Each scalar function has been designed by applying the mesh-based approach with interpolating constraints at (a,b) 2 maxima, (c,d) 2 minima, (e,f) 2 maxima and 2 minima, (g,h) 4 saddles. Pictures in (a,c,e,g) show the constrained maxima and minima as well as the saddles of the designed scalar function. Maxima, minima, and saddles are represented as red, blue, and green dots, respectively.

that

$$(\mathbf{Lf})_i = l_{ii} f(\mathbf{p}_i) - \sum_{j \in \mathcal{N}(i)} l_{ij} f(\mathbf{p}_j)$$

$$= l_{ii} f(\mathbf{p}_i) - \sum_{j \in \mathcal{N}(i) \cap \mathcal{I}^C} l_{ij} f(\mathbf{p}_j) - \sum_{j \in \mathcal{N}(i) \cap \mathcal{I}} l_{ij} f_j.$$

Indicating with $\mathbf{g} := (f(\mathbf{p}_i))_{i \in \mathcal{I}^C} \in \mathbb{R}^{n-k}$ the set of unknowns, the previous identities can be written in matrix form as $\tilde{\mathbf{L}}\mathbf{g} - \mathbf{b}$. Here, $\tilde{\mathbf{L}} \in \mathbb{R}^{(n-k) \times (n-k)}$ is the matrix achieved by cancelling the ith-row and i^{th}-column of $\mathbf{L}$, $i \in \mathcal{I}$, and the entries of the constant term $\mathbf{b} \in \mathbb{R}^{n-k}$ are given by $\sum_{j \in \mathcal{N}(i) \cap \mathcal{I}} l_{ij} f_j$, $i \in \mathcal{I}^C$. Therefore, the constrained least-squares minimization problem (5.3) is equivalent to $\min_{\mathbf{x} \in \mathbb{R}^{n-k}} \{\|\tilde{\mathbf{L}}\mathbf{x} - \mathbf{b}\|_2\}$, where the solution $\mathbf{x}$ provides the f-values at the points associated with $\mathcal{I}^C$. Since the rank of $\mathbf{L}$ is $n - 1$, the rank of $\tilde{\mathbf{L}}$ is $n - k$, $k \geq 1$ and the vector $\mathbf{x}$ is the unique solution to the sparse linear system $\tilde{\mathbf{L}}\mathbf{x} = \mathbf{b}$. In Fig. 5.2(e,f), (g,h), the selection of critical points at symmetric locations makes the behavior of the resulting scalar function symmetric.

Design of scalar functions with least-squares constraints on critical points The function $f : \mathcal{M} \to \mathbb{R}$, which is the best compromise between the least-squares constraint $\sum_{i \in \mathcal{I}} |f(\mathbf{p}_i) - f_i|^2$ and the smoothness term $\|\mathbf{L}\mathbf{f}\|_2$, is defined as the solution to the problem (Fig. 5.3)

$$\min_{\mathbf{f} \in \mathbb{R}^n} \{\mathcal{F}(\mathbf{f})\}, \qquad \mathcal{F}(\mathbf{f}) := \epsilon \sum_{i \in \mathcal{I}} |f(\mathbf{p}_i) - f_i|^2 + \|\mathbf{L}\mathbf{f}\|_2^2. \qquad (5.4)$$

Then, the derivative of $\mathcal{F}$ with respect to the unknown $f(\mathbf{p}_k)$ is

$$\begin{cases} \sum_{i,j=1}^n l_{ij} l_{ik} f(\mathbf{p}_j) + \epsilon(f(\mathbf{p}_k) - f_k), & k \in \mathcal{I}, \\ \sum_{i,j=1}^n l_{ij} l_{ik} f(\mathbf{p}_j), & k \in \mathcal{I}^C, \end{cases}$$

which is re-written in matrix form as $(\mathbf{L}^\top \mathbf{L} + \epsilon \Gamma)\mathbf{f} = \epsilon \mathbf{b}$, with

$$\Gamma_{ij} := \begin{cases} 1 & i = j \in \mathcal{I}, \\ 0 & \text{else}, \end{cases} \qquad b_i := \begin{cases} f_i & i \in \mathcal{I}, \\ 0 & i \in \mathcal{I}^C, \end{cases} \qquad \mathbf{b} \in \mathbb{R}^n, \quad \Gamma \in \mathbb{R}^{n \times n}.$$

The coefficient matrix $\mathbf{L}^\top \mathbf{L} + \epsilon \Gamma$, $\epsilon > 0$, is symmetric, sparse, and positive definite; indeed, our problem admits a unique solution. For the selection of ϵ, which represents the trade-off between approximation accuracy and smoothness of the solution, we apply the L-curve criterion [HO93]; an alternative is to consider statistical criteria [Wah90]. For surfaces with a high genus or a high number of designed saddles, it is generally preferable to apply least-squares instead of interpolating constraints in order to reduce the number of additional critical points. For instance, least-squares constraints in Fig. 5.3 are associated with a regular distribution of smooth level-sets, even in case of a large number of selected critical points (Fig. 5.3d).

For the mesh-based design of a scalar function with constrained critical points, we can also consider the Laplacian matrix $\tilde{\mathbf{L}} := \mathbf{B}^{-1}\mathbf{L}$ with linear FEM weights [RWSN09], where $\mathbf{L}$ is the Laplacian matrix with cotangent weights and $\mathbf{B}$ is the mass matrix, which encodes the variation of the triangles' area. In this case, we select the norm $\|\mathbf{f}\|_{\mathbf{B}}^2 := \mathbf{f}^\top \mathbf{B} \mathbf{f}$ induced by the symmetric, positive-definite mass matrix $\mathbf{B}$, instead of the $\mathcal{L}^2$-norm used in Eq. (5.3) and Eq. (5.4). The $\mathbf{B}$-norm generally improves the robustness of the resulting scalar function with respect to an irregular sampling of the input surface; in particular, the proposed formulation and its properties remain unchanged. For the design of scalar functions with interpolating constraints, the Dirichlet energy $\|\mathbf{L}\mathbf{f}\|_{\mathbf{B}}$ can be re-written in terms of the $\mathcal{L}^2$-norm as $\|\mathbf{L}\mathbf{f}\|_{\mathbf{B}} = \|\mathbf{B}^{1/2}\mathbf{L}\mathbf{f}\|_2$,

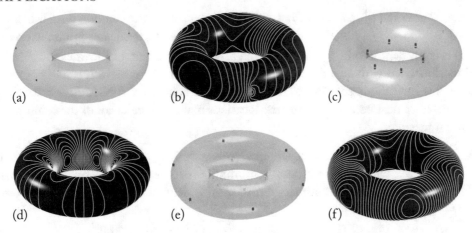

Figure 5.3: (a,c,e) Critical points and (b,d,f) level-sets of the designed scalar function achieved with least-squares constraints (mesh-based approach) at (a,b) 3 maxima, (c,d) 6 saddles (internal circle), (e,f) 3 maxima, 3 minima, and 6 saddles. Pictures in (a,c,e) show the constrained maxima and minima as well as the saddles of the designed scalar function. Maxima, minima, and saddles are represented as red, blue, and green dots, respectively.

where $\mathbf{B}^{1/2}$ is the square-root of $\mathbf{B}$. Then, we can either compute $\mathbf{B}^{1/2}$ [GV89, Ch. 4] or lump $\mathbf{B}$ to the diagonal matrix whose entries are the areas of the Voronoi regions of the mesh vertices. For the mesh-based design of scalar functions with least-squares constraints, the normal equation becomes $(\mathbf{L}^{\top}\mathbf{B}\mathbf{L} + \epsilon\mathbf{B}\Gamma\mathbf{B})\mathbf{f} = \epsilon\mathbf{B}^{\top}\mathbf{b}$. For our experiments, we have selected the Voronoi-cotangent weights (c.f., Eq. (5.3)).

Properties of the designed scalar functions   Recalling the Euler formula Eq. (5.2), the number of saddles of the designed scalar function is $s = 2(g-1) + m + M$, where g is the genus of the input domain. Combining this relation with the smoothness of the approximation due to the minimization of the Laplacian energy, we expect a low number of additional critical points according to the Euler formula, the smoothness and regular distribution of the level-sets on the input surface. Choosing 1 maximum and 1 minimum, the minimal number of saddles is $s = 2g$. The selection of up-to $2g$ saddles and interpolating constraints, together with the minimization of the $\mathcal{L}^2$-norm (*mesh design*) or the Dirichlet energy (*meshless design*), provide the smoothest scalar function that satisfies the selected conditions on the number and location of the critical points. Selecting more than $2g$ saddles induces additional extrema, which are necessary to guarantee the validity of the Euler formula. Assuming that the input scalar function is general (i.e., $f(\mathbf{p}_i) \neq f(\mathbf{p}_j)$, (i, j) edge), from the Euler formula we get that the additional critical points $\tilde{m}, \tilde{M}, \tilde{s}$, satisfy the "*nullity relation*" $\tilde{m} - \tilde{s} + \tilde{M} = 0$.

5.2   LAPLACIAN SMOOTHING OF SCALAR FUNCTIONS

Scalar functions are extensively used to model data in engineering, geographical applications, scientific visualization, and bio-informatics. In each of these research fields, a variety of phenomena is described by a large set of data modeled as the values of a scalar function defined on a surface. These values can be acquired from the real world (e.g., terrain models in GIS) or generated by solving simulation problems (e.g., fluid dynamics, heat equation [BN03, NGH04]).

In the aforementioned applications, an arbitrary scalar function $f : \mathcal{M} \rightarrow \mathbb{R}$, defined on a 2-manifold surface $\mathcal{M}$, is usually associated with a high differential noise, which is due to a low quality of the discrete representations of the input data, unstable computations, and numerical approximations. Here, as *differential noise* of f we refer to a high number of critical points, which have very close positions or f-values, include multiple saddles, and generally do not verify the Euler formula. From our perspective, the critical points of f are a natural choice to guide the approximation and smoothing of f; in fact, they usually represent relevant information about its behavior. Controlling the distribution of the critical points of smooth approximations of noisy maps is also crucial for quadrilateral remeshing [LYL17, LYCL17, DBG$^+$06, NGH04], and topological analysis [BEHP04, EHNP04, EMP06]. In the following, as (discrete) smooth approximation of f we refer to any approximation of f with regular (i.e., not noisy) level sets and a generally low number of non-clustered critical points.

In this context, we introduce the classification and simplification of the critical points based on persistence (Sec. 5.2.1), the smoothing of noisy scalar function. Then, this discussion will be used to introduce the unconstrained and constrained smoothing (Sec. 5.2.2) of noisy scalar functions.

5.2.1   RELATED WORK ON SMOOTHING

We briefly review previous work on the classification and simplification of the critical points based on persistence, and two main examples of smoothing operators, i.e., the bilateral filtering and the Laplacian smoothing. Then, we discuss the bilateral filtering and the filtered Laplacian smoothing, whose main drawback is the lack of control on the final number and distribution of the critical points of the smoothed function.

Topological simplification of critical points based on persistence   Given a scalar function $f : \mathcal{M} \rightarrow \mathbb{R}$ with a large number of critical points associated with a low variation of the corresponding f-values, previous work [BEHP04, EHNP04] defines a topological hierarchy for f that is constructed by performing a progressive simplification of the Morse complex $\mathcal{F}$ of f through the cancellation of pairs of critical points. Then, the critical points are paired by visiting $\mathcal{M}$ with respect to the reordering of its vertices according to increasing values of f. The importance weight associated with the pair $(\mathbf{p}_i, \mathbf{p}_j)$ is measured as the *persistence* of $\mathbf{p}_i, \mathbf{p}_j$, that is, $|f(\mathbf{p}_i) - f(\mathbf{p}_j)|$. The local updates of the complex are performed by iteratively removing those pairs with the lowest persistence and reconnecting the neighbors of the removed nodes. Each node removal affects the number and configuration of the critical points of $\mathcal{F}$ without changing f. Therefore, the simplification provides a hierarchy for f where each Morse complex $\mathcal{F}^{(k)}$ is not associated with a corresponding scalar function $f^{(k)}$ on $\mathcal{M}$.

In [EMP06], the input scalar function f is replaced with a new function $\tilde{f}$ that has the same points of persistency of f higher than a given threshold ϵ and the $\mathcal{L}_\infty$-error between f and $\tilde{f}$ is lower than ϵ. The ϵ-*simplification* of the structure of f and the construction of $\tilde{f}$ are based on an iterative process, which cancels minimum-saddle pairs by sweeping the vertices from bottom to top and lower the saddles that belong to a pair of persistency lower than ϵ.

To improve the approximation accuracy and the smoothness of the solution, topological simplification is combined with the minimization of the $\mathcal{L}_\infty$ error between the input and the simplified functions [TP12, TGP15], isotropic Laplacian [DBG$^+$06, NGH04, Tau95] or Gaussian [LZ07] filters, and least-squares approximation constrained to preserve the persistent critical points [PF09, TP12].

Bilateral filtering   In [LZ07], a bilateral filter operator updates $f(\mathbf{p}_i)$ using a weighted average $s_\sigma(\mathbf{p}_i)$ of the function differences between its neighboring vertices and $\mathbf{p}_i$. For $i = 1, \ldots, n$, this value is defined as

$$s_\sigma(\mathbf{p}_i) := \frac{\sum_{\mathbf{p}_j \in \mathcal{N}(\mathbf{p}_i, \sigma)} f_{ij} \varphi_{\sigma_1}(d_{ij}) \varphi_{\sigma_2}(f_{ij})}{\sum_{\mathbf{p}_j \in \mathcal{N}(\mathbf{p}_i, \sigma)} \varphi_{\sigma_1}(d_{ij}) \varphi_{\sigma_2}(f_{ij})},$$

with weights $d_{ij} := \|\mathbf{p}_j - \mathbf{p}_i\|_2$ and $f_{ij} := |f(\mathbf{p}_j) - f(\mathbf{p}_i)|$. The function $\varphi_\sigma(t) := e^{-\frac{t^2}{2\sigma}}$ is the Gaussian kernel of support σ and $\mathcal{N}(\mathbf{p}_i, \sigma)$ is the set of vertices of $\mathcal{M}$ that fall inside the sphere of center $\mathbf{p}_i$ and radius σ. Once the value $s_\sigma(\mathbf{p}_i)$ has been computed, each function value $f(\mathbf{p}_i)$ is updated to $\tilde{f}(\mathbf{p}_i) := f(\mathbf{p}_i) + s_\sigma(\mathbf{p}_i)$ and the iteration proceeds until a chosen number k of steps is reached. This approach requires to set the parameters $\sigma_1, \sigma_2, \sigma, k$ and does not have a direct control on the final number of preserved critical points. The methods in [EHNP04, EMP06, LZ07] take $\mathcal{O}(n \log n)$-time and the Laplacian smoothing [Tau95] is linear in the number n of vertices of $\mathcal{M}$.

Filtered Laplacian smoothing operators   The smoothness of a scalar function $f : \mathcal{M} \to \mathbb{R}$ is defined by imposing that the value of f at a vertex differs as little as possible from the f-values on its neighbors. Therefore, a smoothing operator [Tau95] assigns to each value $f(\mathbf{p}_i)$ the difference between $f(\mathbf{p}_i)$ and the weighted average of its neighbors, that is,

$$f(\mathbf{p}_i) - \frac{\sum_{j \in \mathcal{N}(i)} w_{ij} f(\mathbf{p}_j)}{\sum_{j \in \mathcal{N}(i)} w_{ij}}, \quad i = 1, \ldots, n,$$

with w_{ij} real weight. The coefficients w_{ij} have been computed by minimizing the Dirichlet energy [DMSB99, PP93] and the mean-value theorem [Flo03]. Indeed, we can consider the Laplacian smoothing $\mathbf{L}f$ or select a polynomial *transfer function* φ and define the *Laplacian low-pass filter* $\mathbf{f} \to \varphi(\mathbf{L})\mathbf{f}$ [NGH04, Tau95]. Small powers of $\mathbf{L}$ attenuate higher frequencies of f and the definition of the Laplacian filter resembles the convolution operator. The main drawback of the method is the lack of control on the number and position of the critical points. Indeed, the idea behind the simplification of the critical points proposed in [PF09] is to modify the definition in [Ban67b] by checking the changes of the sign of f along the edges of the 1-star of each vertex with respect to a positive threshold and to apply a constrained least-squares approximation to recover the underlying smooth functions by preserving the simplified critical points. For more details, we refer the reader to Sec. 5.2.2.

5.2.2 UN-CONSTRAINED AND CONSTRAINED LAPLACIAN SMOOTHING OF SCALAR FUNCTIONS

According to [PF09] and adapting Tikhonov regularization [TA77] to scalar functions defined on surfaces, we introduce an unconstrained smoothing algorithm based on the minimization of a functional, which is a trade-off between approximation accuracy and smoothness of the solution. This choice also allows us to easily insert constraints in the smoothing process and to control the number of preserved critical points. The constrained and unconstrained smoothing reduces to solving a sparse linear system with direct or iterative techniques [GV89].

In case of interpolating constraints, the set of critical points of the approximation $\tilde{f}$ of the input function f contains $\mathcal{C}$ plus a number of additional and well-behaved maxima, minima, and saddles, which is low with respect to those of f. With our approach, the points of $\mathcal{C}$ are preserved in $\tilde{f}$ without diffusing them. On the contrary, the isotropy of the Laplacian matrix indiscriminately smooths noise and topological features [DBG$^+$06, NGH04, Tau95] without constraints on their relocations or cancellations. Constrained least-squares techniques [SCOIT05] have been efficiently applied to define compression schemes based on the selection of a set of anchors. While in [SCOIT05] the choice of the constrained vertices is guided by the final approximation accuracy of the reconstructed surface, in this work the emphasis is on the preservation of the differential properties of f through the simplification of its critical points. Figure 5.4 gives an overview of the proposed approach.

Even though we mainly use the critical points to guide its smoothing and approximation, other choices of the set of feature points are possible without changing the overall structure of the proposed approach. For instance, the feature points can be defined through the analysis of the function values based on clustering techniques (e.g., principal component analysis, k-means clustering) or guided by *a-priori* information on the input scalar function or the application context. Finally, the computational cost of the proposed framework is $\mathcal{O}(n \log n)$, where n is the number of vertices of $\mathcal{M}$.

Un-constrained Laplacian Smoothing of Scalar Functions

Let $\mathcal{K} : \mathcal{H} \to \mathbb{R}$ be an operator defined on a linear space $\mathcal{H}$ of functions, e.g., the space of square-integrable functions on a 2-manifold or a Reproducing Kernel Hilbert Space [Aro50]. Tikhonov regularization [Ber86, TA77] is commonly used to transform the ill-posed problem $\mathcal{K}f = g$,

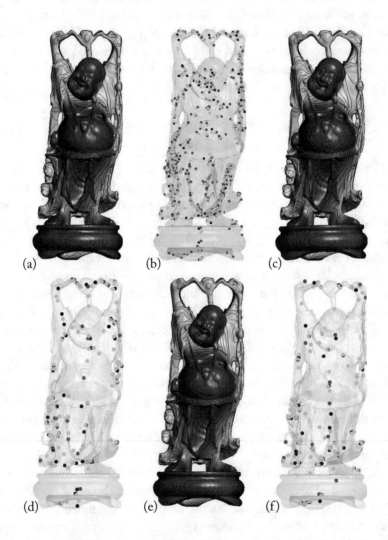

Figure 5.4: Level sets and critical points (a,b) of an input scalar function ($m = 174$, $M = 180$, $s = 370$) and its smoothed version achieved by applying the Tikhonov regularization without (c,d) ($m = 44$, $M = 31$, $s = 91$) and (e,f) with ($m = 65$, $M = 23$, $s = 104$) least-squares constraints. In both cases, the $\mathcal{L}_\infty$-approximation error is lower than 1%. Red, black, and green points locate the m minima, M maxima, and s saddle points of the corresponding function.

$f \in \mathcal{H}$, in a well-posed one; to this end, the regularised solution is computed by minimizing the functional

$$\mathcal{F}(f) := \epsilon \|\mathcal{K}f - g\|_{\mathcal{H}}^2 + \|\mathcal{L}f\|_{\mathcal{H}}^2, \ f \in \mathcal{H},$$

on the linear space $\mathcal{H}$. Here, $\mathcal{L}$ is a regularization operator (e.g., the derivatives of a given order) and ϵ is a strictly positive constant, which defines the trade-off between the approximation error $\|\mathcal{K}f - g\|_{\mathcal{H}}$ and the smoothness energy $\|\mathcal{L}f\|_{\mathcal{H}}$.

In the discrete setting, we apply the Tikhonov regularization to the scalar function $f : \mathcal{M} \to \mathbb{R}$, defined on a triangulated surface $\mathcal{M}$. Therefore, we replace f with the function $\tilde{f} : \mathcal{M} \to \mathbb{R}$, whose values $\tilde{\mathbf{f}} := (\tilde{f}(\mathbf{p}_i))_{i=1}^n \in \mathbb{R}^n$ at the mesh vertices minimize the functional

$$\mathcal{F}(\tilde{\mathbf{f}}) := \epsilon \|\tilde{\mathbf{f}} - \mathbf{f}\|_{\mathbf{B}}^2 + \|\mathbf{L}\tilde{\mathbf{f}}\|_{\mathbf{B}}^2. \tag{5.5}$$

A natural choice of the smoothing operator $\mathbf{L}$ is the Laplacian matrix (Sec. 1.1.1) associated with $\mathcal{M}$. Therefore, $\mathcal{F}(\tilde{\mathbf{f}})$ is defined as the compromise between approximation accuracy $\|\tilde{\mathbf{f}} - \mathbf{f}\|_{\mathbf{B}}$ and smoothness $\|\mathbf{L}\tilde{\mathbf{f}}\|_{\mathbf{B}}$ of the solution. In this case, the normal equations of the new functional $\mathcal{F}(\tilde{\mathbf{f}}) := \epsilon \|\tilde{\mathbf{f}} - \mathbf{f}\|_{\mathbf{B}}^2 + \|\mathbf{L}\tilde{\mathbf{f}}\|_{\mathbf{B}}^2$ is

$$(\mathbf{L}^\top \mathbf{B} \mathbf{L} + \epsilon \mathbf{B})\tilde{\mathbf{f}} = \epsilon \mathbf{B} \mathbf{f}. \tag{5.6}$$

The coefficient matrix in (5.6) is sparse, symmetric, and positive definite; in particular, $\tilde{\mathbf{f}}$ is uniquely defined as solution of the previous linear system (Fig. 5.5). Recalling the relation (1.1), we get that the normal equation is equivalent to

$$\left[\Lambda^2 + \epsilon \mathbf{I}\right] \mathbf{X}^\top \mathbf{B} \tilde{\mathbf{f}} = \epsilon \mathbf{X}^\top \mathbf{B} \mathbf{f}.$$

Using the generalized Laplacian eigensystem, we express $\tilde{\mathbf{f}} = \sum_{i=1}^n \alpha_i \mathbf{x}_i$ as a linear combination of the eigenvectors and compute the corresponding coefficients as $\left[\Lambda^2 + \epsilon \mathbf{I}\right] \alpha = \epsilon \mathbf{X}^\top \mathbf{B} \mathbf{f}$, i.e.,

$$\alpha = \epsilon \left[\Lambda^2 + \epsilon \mathbf{I}\right]^{-1} \mathbf{X}^\top \mathbf{B} \mathbf{f} = \left(\epsilon (\lambda_i^2 + \epsilon)^{-1} \langle \mathbf{f}, \mathbf{x}_i \rangle_{\mathbf{B}}\right)_{i=1}^n,$$

and the minimum of the functional (5.5) is attained at (Fig. 5.6)

$$\tilde{\mathbf{f}} = \sum_{i=1}^n \frac{\epsilon}{\lambda_i^2 + \epsilon} \langle \mathbf{f}, \mathbf{x}_i \rangle_{\mathbf{B}} \mathbf{x}_i. \tag{5.7}$$

Assuming that the noise in f is concentrated in high frequency and the Fourier coefficient $\langle \mathbf{f}, \mathbf{x}_i \rangle_{\mathbf{B}}$ decays rapidly with i, most of the energy of f can be reconstructed from the lower frequency components $\mathbf{x}_i$ and the noise contribution along $\mathbf{x}_i$ is small. From Eq. (5.7), we conclude that the regularization term $(\lambda_i^2 + \epsilon)^{-1}$ filters out the contributions to the solution corresponding to the high eigenvalues (Fig. 5.7).

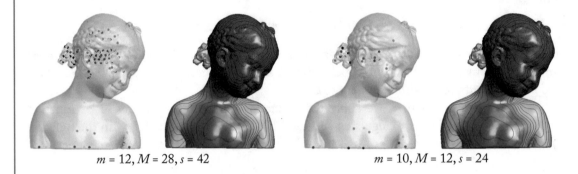

$$m = 12, M = 28, s = 42 \qquad\qquad m = 10, M = 12, s = 24$$

Figure 5.5: Tikhonov regularization without constraints and achieved using the linear FEM approximation of the Laplacian matrix. The level sets and critical points of the input and smoothed scalar function are shown in the first and second row, respectively. The threshold and the approximation error are $\epsilon := 0.1$, $\mathcal{L}_\infty = 0.081$, respectively.

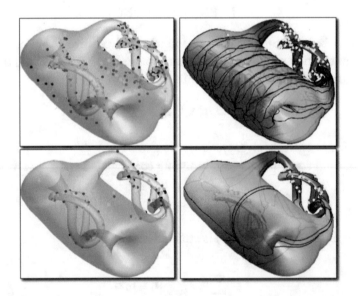

Figure 5.6: *First row*: critical points and Morse complex of a noisy scalar function f ($M = 18$, $m = 38$, $s = 58$). *Second row*: critical points of the scalar function $\tilde{f}$ ($m = 10$, $M = 12$, $s = 24$) achieved by smoothing f with least-squares constraints on the set of preserved critical points shown in (a, right corner).

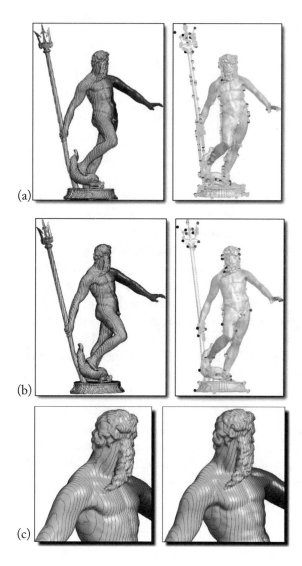

Figure 5.7: (a) Noisy ($m = 127$, $M = 57$, $s = 188$) and (b) smoothed scalar function ($m = 12$, $M = 14$, $s = 30$). (c) Zoom-in on the level sets of the input (left) and smoothed (right) function.

Error bounds for the smoothed scalar function   Let us suppose that we perturb the f-values and consider the new scalar function $f_e : \mathcal{M} \to \mathbb{R}$ whose values on the vertices of $\mathcal{M}$ are $\mathbf{f}_e := (f(\mathbf{p}_i) + e_i)_{i=1}^n$, with $\mathbf{e} := (e_i)_{i=1}^n$ perturbation vector. Then, we denote with $\tilde{\mathbf{f}}$ and $\tilde{\mathbf{f}}_e$ the solution to the unperturbed and perturbed problem, respectively. Indeed, the variation

$$
\begin{aligned}
\|\tilde{\mathbf{f}}_e - \tilde{\mathbf{f}}\|_{\mathbf{B}}^2 &= \left\| \sum_{i=1}^n \frac{\epsilon}{\lambda_i^2 + \epsilon} \langle \mathbf{e}, \mathbf{x}_i \rangle_{\mathbf{B}} \mathbf{x}_i \right\|_{\mathbf{B}}^2 \\
&= \sum_{i=1}^n \frac{\epsilon^2}{(\lambda_i^2 + \epsilon)^2} |\langle \mathbf{e}, \mathbf{x}_i \rangle_{\mathbf{B}}|^2 \\
&\leq \epsilon^2 \|\mathbf{e}\|_{\mathbf{B}}^2 \sum_{i=1}^n \frac{1}{(\lambda_i^2 + \epsilon)^2} \\
&\leq \epsilon^2 C \|\mathbf{e}\|_{\mathbf{B}}^2, \qquad C := \int_0^{+\infty} (s^2 + \epsilon)^{-2} ds,
\end{aligned}
$$

is bounded by the **B**-norm of the perturbation error.

In the approximation schemes previously described, as ϵ increases (Fig. 5.8), the approximation error dominates the value of the functional $\mathcal{F}$ in Eq. (5.5); therefore, the solution is forced to precisely approximate all the f-values on $\mathcal{M}$ and the error $\|\mathbf{f} - \tilde{\mathbf{f}}\|_{\mathbf{B}}$ is minimized. In particular, the number of critical points of $\tilde{f}$ increases with ϵ. As ϵ tends to zero, the smoothness of the approximation $\tilde{f}$ of f becomes predominant and filters out close critical points and/or f-values. In this case, we have a lower number of critical points and a higher approximation error, which is estimated according to the previous upper bounds. A detailed discussion on the choice of ϵ is presented in [HO93, Wah90]. Figures 5.7 and 5.9 show the selection of the optimal value of the threshold ϵ, which is based on the L-curve criterion [HO93] and provides the approximation of f that is the best compromise between accuracy and smoothness.

It is worth noting that these schemes do not take into account the configuration of the critical points of f and are guided by all the f-values at the mesh vertices. Therefore, in the following paragraph we change Eq. (5.5) in such a way that the solution of the corresponding minimization problem is forced to accurately approximate those f-values that characterize the behavior of f.

Smoothing Functions with Constrained Critical Points

In several cases, the critical points of a scalar function $f : \mathcal{M} \to \mathbb{R}$, defined on a 3D shape $\mathcal{M}$, are more essential than geometric error to analyze the properties measured of f and $\mathcal{M}$. For instance, the spectral quadrilateral remeshing [DBG$^+$06, HZM$^+$08] is mainly guided by the number and position of the critical points of the Laplacian eigenfunctions. Therefore, smoothing techniques constrained to preserve a subset of the critical points of an input scalar function, while discarding redundant critical points with close positions and function values, provide a flexible control over the number, shape, and size of the resulting quadrangular patches. Remov-

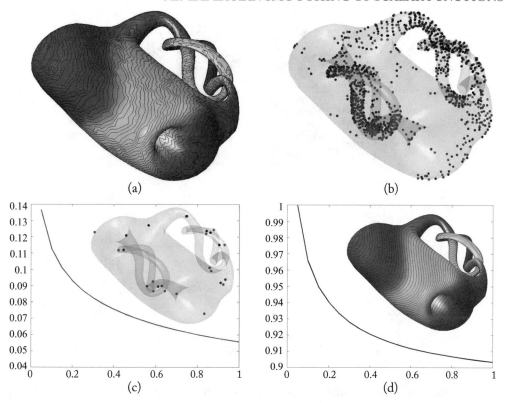

Figure 5.8: (a) Level sets and (b) critical points of a noisy scalar function f ($m = 312$, $M = 280$, $s = 594$). Variation (y-axis) of the (c) $\mathcal{L}_\infty$-approximation error and (d) the Sobolev semi-norm (Sec. 5.2.1) of the Tikhonov approximation of f with respect to several thresholds ϵ (x-axis). The pictures (c,d) also show the function that represents the best compromise between approximation accuracy and smoothness.

ing clustered critical points and filtering small variations of the function values also diminish the number of patches and improves the smoothness of the patch boundaries.

Our aim is to define an approximation scheme that preserves the topological features of $f : \mathcal{M} \to \mathbb{R}$ through its critical points, or a subset of them, and distinguishes the global structure of f from its local details. To this end, we compute a smooth approximation $\tilde{f} : \mathcal{M} \to \mathbb{R}$ of f that preserves/approximates a subset of the critical points of f. More precisely, let $\{\mathbf{p}_i, i \in \mathcal{C}\}$ be the set of critical points of f or those that have been preserved after the simplification of low persistent critical points. Then, we consider the set

$$\mathcal{I} := \mathcal{C} \cup \{j \in \mathcal{N}(i), i \in \mathcal{C}\},$$

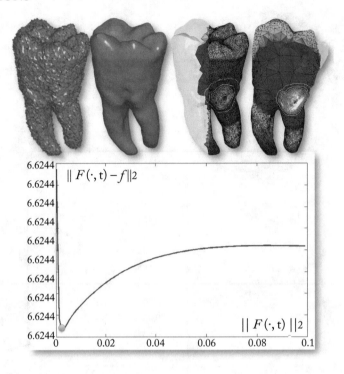

Figure 5.9: Selection of the optimal scale and corresponding volumetric diffusion smoothing (upper part, right), Padé-Chebyshev approximation of degree $r = 7$) on the noisy volumetric model of the teeth (upper part, left).

whose indices belong to $\mathcal{C}$ and to the corresponding 1-stars $\{\mathcal{N}(i),\ i \in \mathcal{C}\}$. Assuming that the indices in $\mathcal{I}$ are without repetitions and that its cardinality is k, we compute the approximation $\tilde{f}$ of f using the set $\{f(\mathbf{p}_i),\ i \in \mathcal{I}\}$ as interpolating constraints. To this end, we define $\tilde{f}$ as the solution of the constrained minimization problem in Eq. (5.3). In this way, we have that the set $\tilde{\mathcal{C}}$ of critical points of $\tilde{f}$ contains the set $\mathcal{C}$. To compute the solution of the aforementioned problem, we follow the approach described in Sec. 5.1 (c.f., Eq. (5.2)). Examples are shown in Fig. 5.10 and Fig. 5.11.

Laplacian and diffusion basis functions   Even though the Laplacian eigenvectors are intrinsic to the input surface, they can be computed only for a small set of eigenvalues and do not provide a flexible alignment of the function behavior to specific shape features. The geometry-aware functions [SCOIT05] provide a computationally efficient way to encode the local geometric information of $\mathcal{M}$; and a similar approach can be applied to define more general classes of basis functions on a given shape. Applying the heat kernel matrix, we can define the *diffusion basis* $\mathcal{B} := \{\mathbf{K}_t \mathbf{e}_i\}_{i=1}^n$, whose elements have a smooth behavior on $\mathcal{M}$ and are intrinsically

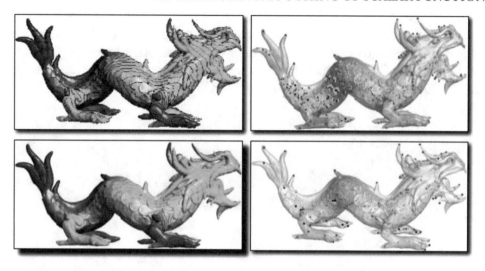

Figure 5.10: Tikhonov smoothing with interpolating constraints. The level sets and critical points of the input and smoothed scalar function are shown in the first and second row. The $\mathcal{L}_\infty$-error is 0.08.

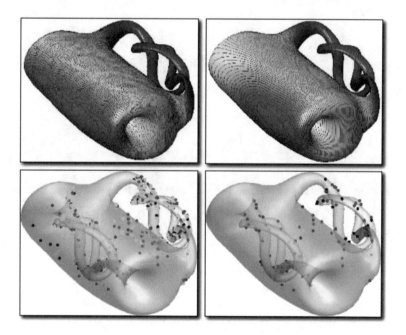

Figure 5.11: *First row*: level sets of the noisy f (left) and smoothed $\tilde{f}$ (right) scalar function. *Second row*: critical points of (left) f ($M = 18, m = 38, s = 58$) and (right) simplified set ($M = 10, m = 12, s = 10$). This last set of critical points are used as interpolating constraints to compute $\tilde{f}$ ($M = 12, m = 14, s = 28$). See also Fig. 5.6.

defined by $\mathcal{M}$ (Fig. 5.12). To define a set of *shape-driven* canonical basis functions, as feature points $\{\mathbf{p}_i\}_{i \in \mathcal{A}}$ of a 3D shape we select the maxima and minima of the Laplacian eigenfunctions related to the smallest eigenvalues [RPSS10] or of the auto-diffusion functions [GBAL09]. Finally, the definition of different basis function is also fundamental to define functions between shapes [GCO06, GMGP05, HK03, LG05, MS05, OFCD02, OBCS+12, RPSS10].

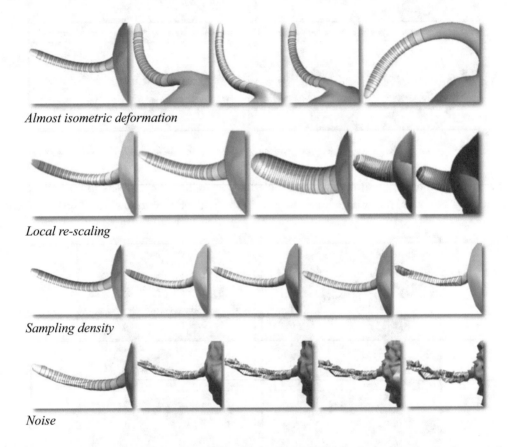

Almost isometric deformation

Local re-scaling

Sampling density

Noise

Figure 5.12: Robustness of the computation of the linear FEM heat kernel from a seed point placed on the spike of the tail. The transformation strength increases from left to right.

Smoothing medical data   In medical applications, the heat kernel is central in diffusion filtering and smoothing of images [ALM92, FCC92, PM90, SKS07, TWBO02, Wit83], 3D shapes [BX02, GSS99], and anatomical surfaces [CRD+05, KCS+12, WZS+13, Pat15]. Figure 2.1 compares the diffusion smoothing of a noisy data set computed with the Padé-Chebyshev approximation of degree $r = 7$ and the truncated approximation with k Laplacian eigenparis. A low number of eigenpairs does not preserve shape details; increasing k reconstructs the surface noise. The ℓ_∞ error between (a) and the smooth approximation of (b) is lower than 1% for (c) the Padé-Chebyshev method, and (d) varies from 12% ($k = 100$) up to 13% ($k = 1K$) for the truncated spectral approximation.

CHAPTER 6

Conclusions

This book provides a common background on the definition and numerical computation of Laplacian spectral kernels and distances for geometry processing and shape analysis, as a generalization of the well-known biharmonic, diffusion, and wave distances. To support the reader in the selection of the most appropriate with respect to shape representation, computational resources, and target application, all the reviewed numerical schemes have been discussed and compared in terms of robustness, approximation accuracy, and computational cost.

From the numerical point of view, the evaluation of full shape descriptors (e.g., heat kernel values among all the input points) is partially limited in case of densely sampled shapes, due to the expensive computational time and storage overhead. Indeed, the appropriate selection of seed points on the input domain and the conversion of the spectral descriptor to a sparse approximation are still crucial steps for the evaluation of full shape descriptors. Furthermore, the robustness of the spectrum-free computation to sampling and missing parts suggests the use of the spectral distances and descriptor for partial shape matching.

From the point of view of the definition of the spectral kernels and distances, we have discussed the general properties of the filter that guarantee their well-posedness, intrinsic and invariance properties. A deeper analysis of the filter properties with respect to the induced spectral distances would be beneficial to improve current results on surface watermarking and shape comparison. Finally, learning the filter from the geometric properties of a given class of data is an efficient, but partially unexplored, way to address shape segmentation, comparison, and more generally manifold learning.

As future work, we mention the computation of the solution to a larger class of differential equations induced by the Laplace-Beltrami operator, such as in-homogeneous diffusion and wave problems, the study of their main properties, and their applications to shape analysis.

Bibliography

[ABBK11] Y. Aflalo, A. M. Bronstein, M. M. Bronstein, and R. Kimmel. Deformable shape retrieval by learning diffusion kernels. In *Scale Space and Variational Methods in Computer Vision*, pages 689–700, 2011. DOI: 10.1007/978-3-642-24785-9_58. 41, 42, 43

[ACSYD05] P. Alliez, D. Cohen-Steiner, M. Yvinec, and M. Desbrun. Variational tetrahedral meshing. *ACM Transactions on Graphics*, 24(3):617–625, 2005. DOI: 10.1145/1073204.1073238. 4

[AKY99] C. J. Alpert, A. B. Kahng, and S.-Z. Yao. Spectral partitioning with multiple eigenvectors. *Discrete Applied Mathematics*, 90(1–3):3–26, 1999. DOI: 10.1016/s0166-218x(98)00083-3. 8

[All07] G. G. Allaire. *Numerical Analysis and Optimization: An Introduction to Mathematical Modelling and Numerical Simulation*. Numerical Mathematics and Scientific Computation. Oxford University Press, Incorporated, 2007. 7, 15, 23

[ALM92] L. Alvarez, P.-L. Lions, and J.-M. Morel. Image selective smoothing and edge detection by nonlinear diffusion. *SIAM Journal on Numerical Analysis*, 29(3):845–866, 1992. DOI: 10.1137/0729052. 88

[AMN+98] S. Arya, D. M. Mount, N. S. Netanyahu, R. Silverman, and A. Y. Wu. An optimal algorithm for approximate nearest neighbor searching fixed dimensions. *Journal of the ACM*, 45(6):891–923, 1998. DOI: 10.1145/293347.293348. 4

[ARAC14] M. Andreux, E. Rodola, M. Aubry, and D. Cremers. Anisotropic Laplace-Beltrami operators for shape analysis. In *6th Workshop on Non-rigid Shape Analysis and Deformable Image Alignment (NORDIA)*, 2014. DOI: 10.1007/978-3-319-16220-1_21. 1, 3

[Aro50] N. Aronszajn. Theory of reproducing kernels. *Transactions of the American Mathematical Society*, 68:337–404, 1950. DOI: 10.2307/1990404. 33, 79

[ASC11] M. Aubry, U. Schlickewei, and D. Cremers. The wave kernel signature: A quantum mechanical approach to shape analysis. In *IEEE Computer Vision Workshops*, pages 1626–1633, 2011. DOI: 10.1109/iccvw.2011.6130444. xxiii, 9, 42

[AW11]   M. Alexa and M. Wardetzky. Discrete Laplacians on general polygonal meshes. *ACM Transactions on Graphics*, 30(4), 2011. DOI: 10.1145/2010324.1964997. 4

[Ban67a]   T. Banchoff. Critical points and curvature for embedded polyhedra. *Journal of Differential Geometry*, 1:245–256, 1967. DOI: 10.4310/jdg/1214428092. 72

[Ban67b]   T. Banchoff. Critical points and curvature for embedded polyhedra. *Journal of Differential Geometry*, 1:245–256, 1967. DOI: 10.4310/jdg/1214428092. 72, 79

[BB11a]   M. M. Bronstein and A. M. Bronstein. Shape recognition with spectral distances. *IEEE Transactions on Pattern Analysis and Machine Intelligence*, 33(5):1065–1071, 2011. DOI: 10.1109/tpami.2010.210. xxiii, 9, 34, 42, 67

[BB11b]   M.M. Bronstein and A.M. Bronstein. Shape recognition with spectral distances. *IEEE Transactions on Pattern Analysis and Machine Intelligence*, 33(5):1065 –1071, 2011. DOI: 10.1109/tpami.2010.210. 71

[BBB+10]   A. M. Bronstein, M. M. Bronstein, B. Bustos, U. Castellani, M. Crisani, B. Falcidieno, L. J. Guibas, V. Murino I. Kokkinos, I. Isipiran, M. Ovsjanikov, G. Patanè, M. Spagnuolo, and J. Sun. SHREC 2010: Robust feature detection and description benchmark. *Eurographics Workshop on 3D Object Retrieval*, 2010. 17

[BBC+10]   A. M. Bronstein, M. M. Bronstein, U. Castellani, B. Falcidieno, A. Fusiello, A. Godil, L.J. Guibas, I. Kokkinos, Z. Lian, M. Ovsjanikov, G. Patanè, M. Spagnuolo, and R. Toldo. SHREC 2010: Robust large-scale shape retrieval benchmark. *Eurographics Workshop on 3D Object Retrieval*, 2010. 17

[BBG94]   P. Berard, G. Besson, and S. Gallot. Embedding Riemannian manifolds by their heat kernel. *Geometric and Functional Analysis*, 4(4):373–398, 1994. DOI: 10.1007/bf01896401. xxiii

[BBK+10]   A. Bronstein, M. Bronstein, R. Kimmel, M. Mahmoudi, and G. Sapiro. A Gromov-Hausdorff framework with diffusion geometry for topologically-robust non-rigid shape matching. *International Journal of Computer Vision*, 2–3:266–286, 2010. DOI: 10.1007/s11263-009-0301-6. xxiii, 9, 44

[BBOG11]   A. M. Bronstein, M. M. Bronstein, M. Ovsjanikov, and L. J. Guibas. Shape Google: Geometric words and expressions for invariant shape retrieval. *ACM Transactions on Graphics*, 30(1), 2011. DOI: 10.1145/1899404.1899405. xxiii, 9, 44, 67

[BCA12]   M. Bronstein, U. Castellani, and Bronstein A. Diffusion geometry in shape analysis. *Eurographics Tutorial*, 2012. xxiii

[BEHP04] P.-T. Bremer, H. Edelsbrunner, B. Hamann, and V. Pascucci. A topological hierarchy for functions on triangulated surfaces. *IEEE Transactions on Visualization and Computer Graphics*, 10(4):385–396, 2004. DOI: 10.1109/tvcg.2004.3. 77

[BEKB15] D. Boscaini, D. Eynard, D. Kourounis, and M. M. Bronstein. Shape-from-operator: Recovering shapes from intrinsic operators. *Computer Graphics Forum*, 34(2):265–274, 2015. DOI: 10.1111/cgf.12558. 6

[Ben75] J. L. Bentley. Multidimensional binary search trees used for associative searching. *Communications of the ACM*, 18(9):509–517, 1975. DOI: 10.1145/361002.361007. 4

[Ber86] M. Bertero. *Regularization Methods for Linear Inverse Problems*. Inverse Problems, Springer-Verlag, Berlin, 1986. DOI: 10.1007/bfb0072660. 79

[BK10] M. M. Bronstein and I. Kokkinos. Scale-invariant heat kernel signatures for non-rigid shape recognition. In *IEEE Conference on Computer Vision and Pattern Recognition*, pages 1704–1711, 2010. DOI: 10.1109/cvpr.2010.5539838. 17, 44

[BMM+15] D. Boscaini, J. Masci, S. Melzi, M. M. Bronstein, U. Castellani, and P. Vandergheynst. Learning class-specific descriptors for deformable shapes using localized spectral convolutional networks. *Computer Graphics Forum*, 34(5):13–23, 2015. DOI: 10.1111/cgf.12693. 42

[BMR+16] D. Boscaini, J. Masci, E. Rodolà, M. M. Bronstein, and D. Cremers. Anisotropic diffusion descriptors. *Computer Graphics Forum*, in press, 2016. DOI: 10.1111/cgf.12844. xxiii

[BN03] M. Belkin and P. Niyogi. Laplacian eigenmaps for dimensionality reduction and data representation. *Neural Computations*, 15(6):1373–1396, 2003. DOI: 10.1162/089976603321780317. xxiii, 4, 8, 45, 77

[BN06] M. Belkin and P. Niyogi. Convergence of Laplacian eigenmaps. In *Neural Information Processing Systems*, pages 129–136, 2006. 4

[BN08] M. Belkin and P. Niyogi. Towards a theoretical foundation for Laplacian-based manifold methods. *Journal of Computer System Sciences*, 74(8):1289–1308, 2008. DOI: 10.1016/j.jcss.2007.08.006. 4

[BPS93] S. T. Barnard, A. Pothen, and H. D. Simon. A spectral algorithm for envelope reduction of sparse matrices. In *Proc. of the ACM Supercomputing*, pages 493–502, 1993. DOI: 10.1145/169627.169790. 8

[BS07] A. I. Bobenko and B. A. Springborn. A discrete Laplace–Beltrami operator for simplicial surfaces. *Discrete and Computational Geometry*, 38(4):740–756, 2007. DOI: 10.1007/s00454-007-9006-1. 3

[BSW08] M. Belkin, J. Sun, and Y. Wang. Discrete Laplace operator on meshed surfaces. In *Proc. of the 24th Annual Symposium on Computational Geometry*, pages 278–287, 2008. DOI: 10.1145/1377676.1377725. 4

[BSW09] M. Belkin, J. Sun, and Y. Wang. *Constructing Laplace Operator from Point Clouds in* $\mathbb{R}^d$, chapter 112, pages 1031–1040. ACM Press, 2009. DOI: 10.1137/1.9781611973068.112. 4

[BX02] C. L. Bajaj and G. Xu. Anisotropic diffusion of subdivision surfaces and functions on surfaces. *ACM Transactions on Graphics*, 22:4–32, 2002. DOI: 10.1145/588272.588276. 88

[CDR00] U. Clarenz, U. Diewald, and M. Rumpf. Anisotropic geometric diffusion in surface processing. In *IEEE Visualization*, pages 397–405, 2000. DOI: 10.1109/visual.2000.885721. 19

[Chu97] F. R. K. Chung. *Spectral Graph Theory*. American Mathematical Society, 1997. DOI: 10.1090/cbms/092. 3

[CK04] T. Caelli and Serhiy K. An eigenspace projection clustering method for inexact graph matching. *IEEE Transactions on Pattern Analysis and Machine Intelligence*, 26(4):515–519, 2004. DOI: 10.1109/tpami.2004.1265866. 8

[CL06] R. R. Coifman and S. Lafon. Diffusion maps. *Applied and Computational Harmonic Analysis*, 21(1):5–30, 2006. DOI: 10.1016/j.acha.2006.04.006. xxiii, 9, 44

[CLB+09] M. Chuang, L. Luo, B. J. Brown, S. Rusinkiewicz, and M. Kazhdan. Estimating the Laplace-Beltrami operator by restricting 3D functions. In *Proc. of the Symposium on Geometry Processing*, pages 1475–1484, 2009. DOI: 10.1111/j.1467-8659.2009.01524.x. 3

[CMV69] W. J. Cody, G. Meinardus, and R. S. Varga. Chebyshev rational approximations to $\exp(-z)$ in $(0, +\infty)$ and applications to heat-conduction problems. *Journal of Approximation Theory*, 2:50–65, 1969. DOI: 10.1016/0021-9045(69)90030-6. 21

[CPS13] K. Crane, U. Pinkall, and P. Schröder. Robust fairing via conformal curvature flow. *ACM Transactions on Graphics*, 32(4):61, 2013. DOI: 10.1145/2461912.2461986. 15

[CRA+16] L. Cosmo, E. Rodola, A. Albarelli, F. Memoli, and D. Cremers. Consistent partial matching of shape collections via sparse modeling. *Computer Graphics Forum*, in press, 2016. DOI: 10.1111/cgf.12796. xxiii

[CRD+05] M. K. Chung, S. M. Robbins, F. K. M. Dalton, C. R. J. Davidson, A. L. Alex, and A. C. Evans. Cortical thickness analysis in autism with heat kernel smoothing. *NeuroImage*, 25:1256–1265, 2005. DOI: 10.1016/j.neuroimage.2004.12.052. 88

[CRV84] A. J. Carpenter, A. Ruttan, and R. S. Varga. Extended numerical computations on the "1/9" conjecture in rational approximation theory. In *Rational Approximation and Interpolation*, volume 1105 of *Lecture Notes in Mathematics*, pages 383–411. Springer, 1984. DOI: 10.1007/bfb0072427. 21, 50, 58

[CWS03] O. Chapelle, J. Weston, and B. Schölkopf. Cluster kernels for semi-supervised learning. In *Neural Information Processing Systems*, volume 15, pages 585–592, 2003. xxiii, 45

[CWW13] K. Crane, C. Weischedel, and M. Wardetzky. Geodesics in heat: A new approach to computing distance based on heat flow. *ACM Transactions on Graphics*, 32(5):152:1–152:11, October 2013. DOI: 10.1145/2516971.2516977. 34, 46

[DBG+06] S. Dong, P.-T. Bremer, M. Garland, V. Pascucci, and J. C. Hart. Spectral surface quadrangulation. *ACM Siggraph*, pages 1057–1066, 2006. DOI: 10.1145/1179352.1141993. 71, 77, 78, 79, 84

[dGGV08] F. de Goes, S. Goldenstein, and L. Velho. A hierarchical segmentation of articulated bodies. *Computer Graphics Forum*, 27(5):1349–1356, 2008. DOI: 10.1111/j.1467-8659.2008.01274.x. xxiii, 12, 45

[DKG05] S. Dong, S. Kircher, and M. Garland. Harmonic functions for quadrilateral remeshing of arbitrary manifolds. *Computer Aided Geometric Design*, 22(5):392–423, 2005. DOI: 10.1016/j.cagd.2005.04.004. 5, 9

[DLL+10] T. K. Dey, K. Li, C. Luo, P. Ranjan, I. Safa, and Y. Wang. Persistent heat signature for pose-oblivious matching of incomplete models. *Computer Graphics Forum*, 29(5):1545–1554, 2010. DOI: 10.1111/j.1467-8659.2010.01763.x. 41

[DMSB99] M. Desbrun, M. Meyer, P. Schröder, and A. H. Barr. Implicit fairing of irregular meshes using diffusion and curvature flow. In *ACM Siggraph*, pages 317–324, 1999. DOI: 10.1145/311535.311576. 3, 8, 19, 42, 46, 73, 78

[DPS02] J. Dìaz, J. Petit, and M. Serna. A survey of graph layout problems. *ACM Computing Surveys*, 34(3):313–356, 2002. DOI: 10.1145/568522.568523. 8

[DRW10] T. K. Dey, P. Ranjan, and Y. Wang. Convergence, stability, and discrete approximation of Laplace spectra. *ACM Symposium on Discrete Algorithms*, pages 650–663, 2010. DOI: 10.1137/1.9781611973075.54. xxiii, 44

[DS05] T. K. Dey and J. Sun. An adaptive MLS surface for reconstruction with guarantees. In *ACM Symposium on Geometry Processing*, pages 43–52, 2005. 4

[DZM07] R. Dyer, H. Zhang, and T. Möller. Delaunay mesh construction. In *Proc. of Eurographics Symposium on Geometry Processing*, pages 273–282, 2007. 4

[EH09] H. ElGhawalby and E. R. Hancock. Geometric characterizations of graphs using heat kernel embeddings. In *IMA Conference on the Mathematics of Surfaces*, pages 124–142, 2009. DOI: 10.1007/978-3-642-03596-8_8. xxiii

[EHNP04] H. Edelsbrunner, J. Harer, V. Natarajan, and V. Pascucci. Local and global comparison of continuous functions. In *IEEE Visualization*, pages 275–280, 2004. DOI: 10.1109/visual.2004.68. 77, 78

[EMP06] H. Edelsbrunner, D. Morozov, and V. Pascucci. Persistence-sensitive simplification functions on 2-manifolds. In *Proc. of the Symposium on Computational Geometry*, pages 127–134, 2006. DOI: 10.1145/1137856.1137878. 77, 78

[Fal06] R. D. Falgout. An introduction to algebraic multigrid. *Computing in Science and Engineering*, 8(6):24–33, November 2006. DOI: 10.1109/MCSE.2006.105. 7

[FBCM04] C. Fowlkes, S. Belongie, F. Chung, and J. Malik. Spectral grouping using the Nystrom method. *IEEE Transactions on Pattern Analysis and Machine Intelligence*, 26(2):214–225, 2004. DOI: 10.1109/tpami.2004.1262185. 8

[FCC92] J.-M. Morel F. Cattè, P.-L. Lions and T. Coll. Image selective smoothing and edge detection by nonlinear diffusion. *SIAM Journal on Numerical Analysis*, 29(1):182–193, 1992. DOI: 10.1137/0729012. 88

[Fie73] M. Fiedler. Algebraic connectivity of graphs. *Czechoslovak Mathematical Journal*, 23(98):298–305, 1973. 8, 12

[Flo03] M. S. Floater. Mean value coordinates. *Computer Aided Geometric Design*, 20(1):19–27, 2003. DOI: 10.1016/s0167-8396(03)00002-5. 3, 78

[FPS05] F. Fouss, A. Pirotte, and M. Saerens. A novel way of computing similarities between nodes of a graph, with application to collaborative recommendation. In *IEEE/WIC/ACM International Conference on Web Intelligence*, pages 550–556, 2005. DOI: 10.1109/wi.2005.9. xxiii, 42

[FW12] P. Feng and J. Warren. Discrete bi-Laplacians and biharmonic b-splines. *ACM Transactions on Graphics*, 31(4):115:1–115:11, July 2012. DOI: 10.1145/2185520.2335466. 5

[GBAL09] K. Gebal, J. A. Bærentzen, H. Aanæs, and R. Larsen. Shape analysis using the auto diffusion function. *Computer Graphics Forum*, 28(5):1405–1413, 2009. DOI: 10.1111/j.1467-8659.2009.01517.x. xxiii, 9, 44, 88

[GCO06] R. Gal and D. Cohen-Or. Salient geometric features for partial shape matching and similarity. *ACM Transactions on Graphics*, 25(1):130–150, 2006. DOI: 10.1145/1122501.1122507. 88

[GGS03] C. Gotsman, X. Gu, and A. Sheffer. Fundamentals of spherical parameterization for 3D meshes. In *ACM Siggraph*, pages 358–363, 2003. DOI: 10.1145/1201775.882276. 8

[GK06] E. Gine and V. Koltchinskii. Empirical graph Laplacian approximation of Laplace-Beltrami operators: Large sample results. *IMS Lecture Notes Monograph Series*, 51:238, 2006. DOI: 10.1214/074921706000000888. xxiii

[GMGP05] N. Gelfand, N. J. Mitra, L. J. Guibas, and H. Pottmann. Robust global registration. In *Proc. of the Symposium on Geometry Processing*, page 197, 2005. 88

[Gri06] A. Grigoryan. Heat kernels on weighted manifolds and applications. *Contemporary Mathematics*, (398):93–191, 2006. DOI: 10.1090/conm/398/07486. 11

[GS02] C. S. Gordon and Z. I. Szabo. Isospectral deformations of negatively curved Riemannian manifolds with boundary which are not locally isometric. *Duke Mathematical Journal*, 113(2):355–383, June 2002. DOI: 10.1215/S0012-7094-02-11326-X. 6

[GSS99] I. Guskov, W. Sweldens, and P. Schröder. Multiresolution signal processing for meshes. *ACM Siggraph*, pages 325–334, 1999. DOI: 10.1145/311535.311577. 88

[GV89] G. Golub and G. F. VanLoan. *Matrix Computations*, 2nd ed., John Hopkins University Press, 1989. 9, 17, 20, 21, 22, 50, 57, 58, 59, 76, 79

[HAvL05] M. Hein, J.-Y. Audibert, and U. von Luxburg. From graphs to manifolds—weak and strong pointwise consistency of graph Laplacians. In *Learning Theory*, volume 3559 of *Lecture Notes in Computer Science*, pages 470–485. Springer, 2005. DOI: 10.1007/11503415_32. xxiii, 45

[HDL16] M. Hajij, T. Dey, and X. Li. Segmenting a surface mesh into pants using morse theory. *Graphical Models*, 88:12–21, 2016. DOI: 10.1016/j.gmod.2016.09.003. 8

[HK03]   A. B. Hamza and H. Krim. Geodesic object representation and recognition. In *International Conference on Discrete Geometry for Computer Imagery*, pages 378–387, 2003. DOI: 10.1007/978-3-540-39966-7_36. 88

[HKA15]   P. Herholz, J. E. Kyprianidis, and M. Alexa.   Perfect Laplacians for polygon meshes.   *Computer Graphics Forum*, 34(5):211–218, 2015. DOI: 10.1111/cgf.12709. 4

[HO93]   P. C. Hansen and D. P. O'Leary. The use of the l-curve in the regularization of discrete ill-posed problems. *SIAM Journal of Scientific Computing*, 14(6):1487–1503, 1993. DOI: 10.1137/0914086. 13, 75, 84

[Hoe68]   L. Hoermander. The spectral function of an elliptic operator. *Acta Mathematica*, 121(1):193–218, 1968. DOI: 10.1007/bf02391913. 35, 38

[HPW06]   K. Hildebrandt, K. Polthier, and M. Wardetzky. On the convergence of metric and geometric properties of polyhedral surfaces. *Geometria Dedicata*, pages 89–112, 2006. DOI: 10.1007/s10711-006-9109-5. 3, 9

[HQ12]   T. Hou and H. Qin.   Continuous and discrete Mexican hat wavelet transforms on manifolds.   *Graphical Models*, 74(4):221–232, 2012. DOI: 10.1016/j.gmod.2012.04.010. 42

[HVG11]   D. K. Hammond, P. Vandergheynst, and R. Gribonval. Wavelets on graphs via spectral graph theory. *Applied and Computational Harmonic Analysis*, 30(2):129–150, 2011. DOI: 10.1016/j.acha.2010.04.005. xvi, 6, 41, 42, 44

[HZM+08]   J. Huang, M. Zhang, J. Ma, X. Liu, L. Kobbelt, and H. Bao. Spectral quadrangulation with orientation and alignment control. *ACM Transactions on Graphics*, 27(5):1–9, 2008. DOI: 10.1145/1409060.1409100. 71, 84

[JBPS14]   A. Jacobson, I. Baran, J. Popovic, and O. Sorkine-Hornung. Bounded biharmonic weights for real-time deformation. *Communication of the ACM*, 57(4):99–106, 2014. DOI: 10.1145/1964921.1964973. 5

[JMD+07]   P. Joshi, M. Meyer, T. DeRose, B. Green, and T. Sanocki. Harmonic coordinates for character articulation. *ACM Transactions on Graphics*, 26(3), July 2007. DOI: 10.1145/1239451.1239522. 5

[JZ07]   V. Jain and H. Zhang.   A spectral approach to shape-based retrieval of articulated 3D models. *Computer Aided Design*, 39:398–407, 2007. DOI: 10.1016/j.cad.2007.02.009. 8, 9

[JZvK07] V. Jain, H. Zhang, and O. van Kaick. Non-rigid spectral correspondence of triangle meshes. *International Journal on Shape Modeling*, 13(1):101–124, 2007. DOI: 10.1142/s0218654307000968. 8

[KBB⁺13] A. Kovnatsky, M. M. Bronstein, A. M. Bronstein, K. Glashoff, and R. Kimmel. Coupled quasi-harmonic bases. *Computer Graphics Forum*, 32(2):439–448, 2013. DOI: 10.1111/cgf.12064. 5

[KCS⁺12] S.-G. Kim, M. K. Chung, S. Seo, S. M. Schaefer, C. M. van Reekum, and R. J. Davidson. Heat kernel smoothing via Laplace-Beltrami eigenfunctions and its application to subcortical structure modeling. In *Proc. of Pacific Conference on Advances in Image and Video Technology*, pages 36–47, 2012. DOI: 10.1007/978-3-642-25367-6_4. 88

[KCVS98] L. Kobbelt, S. Campagna, J. Vorsatz, and H.-P. Seidel. Interactive multi-resolution modeling on arbitrary meshes. In *ACM Siggraph*, pages 105–114, 1998. DOI: 10.1145/280814.280831. 8

[KFS13] D. Krishnan, R. Fattal, and R. Szeliski. Efficient preconditioning of Laplacian matrices for computer graphics. *ACM Transactions on Graphics*, 32(4):142:1–142:15, July 2013. DOI: 10.1145/2461912.2461992. 8, 59

[KG00] Z. Karni and C. Gotsman. Spectral compression of mesh geometry. In *ACM Siggraph*, pages 279–286, 2000. DOI: 10.1145/344779.344924. xxiii, 1, 8

[Kor03] Y. Koren. On spectral graph drawing. In *Lecture Notes in Computer Science*, volume 2697, pages 496–508. Springer Verlag, 2003. DOI: 10.1007/3-540-45071-8_50. 8

[KR05] B. Kim and J. Rossignac. Geofilter: Geometric selection of mesh filter parameters. *Computer Graphics Forum*, 24(3):295–302, 2005. DOI: 10.1111/j.1467-8659.2005.00854.x. 8, 60

[KSB12] M. Kazhdan, J. Solomon, and M. Ben-Chen. Can mean-curvature flow be modified to be non-singular? *Computer Graphics Forum*, 31(5):1745–1754, 2012. DOI: 10.1111/j.1467-8659.2012.03179.x. 15

[KTT13] K. I. Kim, J. Tompkin, and C. Theobalt. Curvature-aware regularization on Riemannian submanifolds. In *IEEE International Conference on Computer Vision*, pages 881–888, 2013. DOI: 10.1109/iccv.2013.114. 1

[LBB11] R. Litman, A. M. Bronstein, and M. M. Bronstein. Diffusion-geometric maximally stable component detection in deformable shapes. *Computers & Graphics*, 35(3):549–560, 2011. DOI: 10.1016/j.cag.2011.03.011. 23

[LBB12] R. Litman, A. M. Bronstein, and M. M. Bronstein. Stable volumetric features in deformable shapes. *Computers & Graphics*, 36(5):569–576, 2012. DOI: 10.1016/j.cag.2012.03.034. 23

[LD08] W.-J. Lee and R. P. Duin. An inexact graph comparison approach in joint eigenspace. In *Proc of the International Workshop on Structural, Syntactic, and Statistical Pattern Recognition*, pages 35–44. Springer-Verlag, 2008. DOI: 10.1007/978-3-540-89689-0_8. 8

[Lev06] B. Levy. Laplace-Beltrami eigenfunctions: Towards an algorithm that understands geometry. In *Proc. of Shape Modeling and Applications*, page 13, 2006. DOI: 10.1109/smi.2006.21. xxiii

[LG05] X. Li and I. Guskov. Multi-scale features for approximate alignment of point-based surfaces. In *Proc. of the Symposium on Geometry Processing*, pages 217–226, 2005. DOI: 10.2312/SGP/SGP05/217-226. 88

[LGQ09] X. Li, X. Gu, and H. Qin. Surface mapping using consistent pants decomposition. *IEEE Transactions on Visualization and Computer Graphics*, 15(4), 2009. DOI: 10.1109/tvcg.2008.200. 5

[LGW+07] X. Li, X. Guo, H. Wang, Y. He, D. Gu, and H. Qin. Harmonic volumetric mapping for solid modeling applications. In *Proc. of ACM Symposium on Solid and Physical Modeling*, pages 109–120, 2007. DOI: 10.1145/1236246.1236263. 5

[LI14] X. Li and S. S. Iyengar. On computing mapping of 3D objects: A survey. *ACM Computing Surveys*, 47(2):34:1–34:45, December 2014. DOI: 10.1145/2668020. 5

[LKC06] S. Lafon, Y. Keller, and R. R. Coifman. Data fusion and multicue data matching by diffusion maps. *IEEE Transactions on Pattern Analysis Machine Intelligence*, 28(11):1784–1797, 2006. DOI: 10.1109/tpami.2006.223. xxiii, 9, 44

[LPG12] Y. Liu, B. Prabhakaran, and X. Guo. Point-based manifold harmonics. *IEEE Transactions on Visualization and Computer Graphics*, 18(10):1693–1703, 2012. DOI: 10.1109/tvcg.2011.152. 4

[LRF10] Y. Lipman, R. M. Rustamov, and Thomas A. Funkhouser. Biharmonic distance. *ACM Transactions on Graphics*, 29(3), 2010. DOI: 10.1145/1805964.1805971. xxiii, 5, 46

[LS96] R. B. Lehoucq and D. C. Sorensen. Deflation techniques for an implicitly re-started Arnoldi iteration. *SIAM Journal of Matrix Analysis and Applications*, 17:789–821, 1996. DOI: 10.1137/s0895479895281484. 8, 22, 24, 57

[LSW09] C. Luo, I. Safa, and Y. Wang. Approximating gradients for meshes and point clouds via diffusion metric. *Computer Graphics Forum*, 28:1497–1508(12), 2009. DOI: 10.1111/j.1467-8659.2009.01526.x. xxiii, 8, 9, 45

[LTDZ09] S. Liao, R. Tong, J. Dong, and F. Zhu. Gradient field based inhomogeneous volumetric mesh deformation for maxillofacial surgery simulation. *Computers & Graphics*, 33(3):424–432, 2009. DOI: 10.1016/j.cag.2009.03.018. 4, 23

[LWH03] B. Luo, R. C. Wilson, and E. R. Hancock. Spectral embedding of graphs. *Pattern Recognition*, 36(10):2213–2230, 2003. DOI: 10.1016/s0031-3203(03)00084-0. xxiii

[LXFH15] Y. J. Liu, C. X. Xu, D. Fan, and Y. He. Efficient construction and simplification of Delaunay meshes. *ACM Transactions on Graphics*, 34(6):174:1–174:13, October 2015. DOI: 10.1145/2816795.2818076. 4

[LXH15] Y.-J. Liu, C. X. Xu, and D. F. Ying He. Constructing intrinsic Delaunay triangulations from the dual of geodesic Voronoi diagrams. *CoRR*, abs/1505.05590, 2015. DOI: 10.1145/2999532. 4

[LXW+10] X. Li, H. Xu, S. Wan, Z. Yin, and W. Yu. Feature-aligned harmonic volumetric mapping using MFS. *Computers & Graphics*, 34(3):242–251, 2010. DOI: 10.1016/j.cag.2010.03.004. 5, 23

[LYCL17] C. Liu, W. Yu, Z. Chen, and X. Li. Distributed poly-square mapping for large-scale semi-structured quad mesh generation. *Computer-Aided Design*, 2017. DOI: 10.1016/j.cad.2017.05.005. 77

[LYL17] X. Li, W. Yu, and C. Liu. Geometry-aware partitioning of complex domains for parallel quad meshing. *Computer-Aided Design*, 85:20–33, 2017. 24th International Meshing Roundtable Special Issue: Advances in Mesh Generation. DOI: 10.1016/j.cad.2016.07.014. 77

[LZ07] R. Liu and H. Zhang. Mesh segmentation via spectral embedding and contour analysis. *Eurographics Tutorial*, 26:385–394, 2007. DOI: 10.1111/j.1467-8659.2007.01061.x. 8, 78

[Mal89] J.-L. Mallet. Discrete smooth interpolation. *ACM Transactions on Graphics*, 8(2):121–144, 1989. DOI: 10.1145/62054.62057. 8

[MC10] T. Martin and E. Cohen. Volumetric parameterization of complex objects by respecting multiple materials. *Computers & Graphics*, 34(3):187–197, 2010. DOI: 10.1016/j.cag.2010.03.011. 5

[MCK08] T. Martin, E. Cohen, and M. Kirby. Volumetric parameterization and trivariate B-spline fitting using harmonic functions. In *Proc. of the ACM Symposium on Solid and Physical Modeling*, pages 269–280, 2008. DOI: 10.1145/1364901.1364938. 5, 23

[Mem09] F. Memoli. Spectral Gromov-Wasserstein distances for shape matching. In *Workshop on Non-rigid Shape Analysis and Deformable Image Alignment*, pages 256–263, 2009. DOI: 10.1109/iccvw.2009.5457690. xxiii, 44

[Mem11] F. Memoli. A spectral notion of Gromov-Wasserstein distance and related methods. *Applied and Computational Harmonic Analysis*, 30(3):363–401, 2011. DOI: 10.1016/j.acha.2010.09.005. xxiii, 44

[Mil63] J. Milnor. *Morse Theory*, volume 51 of *Annals of Mathematics Studies*. Princeton University Press, 1963. 72

[MN03] N. J. Mitra and A. Nguyen. Estimating surface normals in noisy point cloud data. In *Proc. of Symposium on Computational Geometry*, pages 322–328. ACM Press, 2003. DOI: 10.1145/777837.777840. 4

[MP93] B. Mohar and S. Poljak. Eigenvalues in combinatorial optimization. *Combinatorial and Graph-theoretical Problems in Linear Algebra*, 23(98):107–151, 1993. DOI: 10.1007/978-1-4613-8354-3_5. 8

[MPSF11] S. Marini, G. Patanè, M. Spagnuolo, and B. Falcidieno. Spectral feature selection for shape characterization and classification. *The Visual Computer*, 27(11):1005–1019, 2011. DOI: 10.1007/s00371-011-0612-9. 8

[MS05] F. Mèmoli and G. Sapiro. A theoretical and computational framework for isometry invariant recognition of point cloud data. *Foundations of Computational Mathematics*, 5(3):313–347, 2005. DOI: 10.1007/s10208-004-0145-y. xxiii, 44, 88

[MS09] M. Mahmoudi and G. Sapiro. Three-dimensional point cloud recognition via distributions of geometric distances. *Graphical Models*, 71(1):22–31, 2009. DOI: 10.1016/j.gmod.2008.10.002. xxiii, 44

[MVL03] C. Moler and C. Van Loan. Nineteen dubious ways to compute the exponential of a matrix, twenty-five years later. *SIAM Review*, 45(1):3–49, 2003. DOI: 10.1137/s00361445024180. 21, 59

[Nad88] N. S. Nadirashvili. Multiple eigenvalue of the Laplace-Beltrami operator. *Mathematics of the USSR-Sbornik*, 61(1):225, 1988. DOI: 10.1070/sm1988v061n01abeh003204. 6

[NGH04] X. Ni, M. Garland, and J. C. Hart. Fair morse functions for extracting the topological structure of a surface mesh. In *ACM Siggraph*, pages 613–622, 2004. DOI: 10.1145/1186562.1015769. 5, 71, 77, 78, 79

[NISA06] A. Nealen, T. Igarashi, O. Sorkine, and M. Alexa. Laplacian mesh optimization. In *Proc. of Computer Graphics and Interactive Techniques*, pages 381–389, 2006. DOI: 10.1145/1174429.1174494. 8

[NJW01] A. Y. Ng, M. I. Jordan, and Y. Weiss. On spectral clustering: Analysis and an algorithm. In *Advances in Neural Information Processing Systems 14*, pages 849–856. MIT Press, 2001. xxiii, 45

[NVT+14] T. Neumann, K. Varanasi, C. Theobalt, M. A. Magnor, and M. Wacker. Compressed manifold modes for mesh processing. *Computer Graphics Forum*, 33(5):35–44, 2014. DOI: 10.1111/cgf.12429. 7

[OBCS+12] M. Ovsjanikov, M. Ben-Chen, J. Solomon, A. Butscher, and L. J. Guibas. Functional maps: A flexible representation of maps between shapes. *ACM Transactions on Graphics*, 31(4):30, 2012. DOI: 10.1145/2185520.2335381. 5, 9, 46, 88

[OFCD02] R. Osada, T. Funkhouser, B. Chazelle, and D. Dobkin. Shape distributions. *ACM Transactions on Graphics*, 21(4):807–832, 2002. DOI: 10.1145/571647.571648. 67, 88

[OMMG10] M. Ovsjanikov, Q. Mèrigot, F. Mèmoli, and L. Guibas. One point isometric matching with the heat kernel. *ACM Symposium on Discrete Algorithms*, pages 650–663, 2010. DOI: 10.1111/j.1467-8659.2010.01764.x. xxiii, 44

[OMT02] R. Ohbuchi, A. Mukaiyama, and S. Takahashi. A frequency-domain approach to watermarking 3D shapes. *Computer Graphics Forum*, 21(3), 2002. DOI: 10.1111/1467-8659.t01-1-00597. 8

[OSV12] L. Orecchia, S. Sachdeva, and N. K. Vishnoi. Approximating the exponential, the Lanczos method and an õ(m)-time spectral algorithm for balanced separator. In *Proc. of the 44th Symposium on Theory of Computing Conference*, pages 1141–1160, 2012. DOI: 10.1145/2213977.2214080. 21

[OTMM01] R. Ohbuchi, S. Takahashi, T. Miyazawa, and A. Mukaiyama. Watermarking 3D polygonal meshes in the mesh spectral domain. In *Graphics Interface*, pages 9–17, 2001. 8

[Pat13] G. Patanè. wFEM heat kernel: Discretization and applications to shape analysis and retrieval. *Computer Aided Geometric Design*, 30(3):276–295, 2013. DOI: 10.1016/j.cagd.2013.01.002. 8, 21, 34

[Pat14]   G. Patanè. Laplacian spectral distances and kernels on 3D shapes. *Pattern Recognition Letters*, 47:102–110, 2014. DOI: 10.1016/j.patrec.2014.04.003. 21, 34, 57

[Pat15]   G. Patanè. Diffusive smoothing of 3D segmented medical data. *Journal of Advanced Research*, 6(3):425–431, 2015. DOI: 10.1016/j.jare.2014.09.003. 88

[Pat16a] G. Patanè. Accurate and efficient computation of Laplacian spectral distances and kernels. *Computer Graphics Forum*, in press, 2016. DOI: 10.1111/cgf.12794. 37, 53, 59, 64

[Pat16b] G. Patanè. STAR—Laplacian spectral kernels and distances for geometry processing and shape analysis. *Computer Graphics Forum*, 35(2):599–624, 2016. DOI: 10.1111/cgf.12866. xxiii

[Pat17a] G. Patanè. Accurate and efficient computation of Laplacian spectral distances and Kernels. *Computer Graphics Forum*, 36(1):184–196, 2017. DOI: 10.1111/cgf.12794. 71

[Pat17b] G. Patanè. Mesh-based and meshless design and approximation of scalar functions. *Computer-aided Geometric Design*, 2017. DOI: 10.1016/j.cagd.2017.05.005. 71

[PF09]   G. Patanè and B. Falcidieno. Computing smooth approximations of scalar functions with constraints. *Computer & Graphics*, 33(3):399–413, 2009. DOI: 10.1016/j.cag.2009.03.014. 8, 9, 78, 79

[PF10]   G. Patanè and B. Falcidieno. Multi-scale feature spaces for shape processing and analysis. In *Proc. of Shape Modeling International*, pages 113–123, 2010. DOI: 10.1109/smi.2010.27. xxiii, 45

[PM90]   P. Perona and J. Malik. Scale-space and edge detection using anisotropic diffusion. *IEEE Transactions on Pattern Analysis and Machine Intelligence*, 12(7):629–639, 1990. DOI: 10.1109/34.56205. 88

[PP93]   U. Pinkall and K. Polthier. Computing discrete minimal surfaces and their conjugates. *Experimental Mathematics*, 2(1):15–36, 1993. DOI: 10.1080/10586458.1993.10504266. 3, 8, 73, 78

[PS12]   G. Patanè and M. Spagnuolo. Local approximation of scalar functions on 3D shapes and volumetric data. *Computers & Graphics*, 36(5):387–397, 2012. DOI: 10.1016/j.cag.2012.03.011. 9

[PS13a]   G. Patanè and M. Spagnuolo. Heat diffusion kernel and distance on surface meshes and point sets. *Computers & Graphics*, 37(6):676–686, 2013. DOI: 10.1016/j.cag.2013.05.019. 8

[PS13b] G. Patanè and M. Spagnuolo. An interactive analysis of harmonic and diffusion equations on discrete 3D shapes. *Computer & Graphics*, 2013. DOI: 10.1016/j.cag.2013.03.006. xxiii, 9

[PSBM07] V. Pascucci, G. Scorzelli, P.-T. Bremer, and A. Mascarenhas. Robust on-line computation of Reeb graphs: Simplicity and speed. In *ACM Siggraph*, pages 58.1–58.9, 2007. DOI: 10.1145/1275808.1276449. 71

[PSF09] G. Patanè, M. Spagnuolo, and B. Falcidieno. A minimal contouring approach to the computation of the reeb graph. *IEEE Transactions on Visualization and Computer Graphics*, 15(4):583–595, 2009. DOI: 10.1109/tvcg.2009.22. 71

[Pus11] M. Pusa. Rational approximations to the matrix exponential in burnup calculations. *Nuclear Science and Engineering*, 169(2):155–167, 2011. DOI: 10.13182/nse10-81. 22

[RBBK10] D. Raviv, M. M. Bronstein, A. M. Bronstein, and R. Kimmel. Volumetric heat kernel signatures. In *Proc. of the ACM Workshop on 3D Object Retrieval*, 3D OR'10, pages 39–44, 2010. DOI: 10.1145/1877808.1877817. xxiii, 23, 44

[RBG+09] M. Reuter, S. Biasotti, D. Giorgi, G. Patanè, and M. Spagnuolo. Discrete Laplace-Beltrami operators for shape analysis and segmentation. *Computer & Graphics*, 33(3):381–390, 2009. DOI: 10.1016/j.cag.2009.03.005. 8, 24

[RCB+16] E. Rodola, L. Cosmo, M. M. Bronstein, A. Torsello, and D. Cremers. Partial functional correspondence. *Computer Graphics Forum*, in press, 2016. DOI: 10.1111/cgf.12797. xxiii

[Ros97] S. Rosenberg. *The Laplacian on a Riemannian Manifold*. Cambridge University Press, 1997. DOI: 10.1017/cbo9780511623783.002. 1, 5

[RPSS10] M. R. Ruggeri, G. Patanè, M. Spagnuolo, and D. Saupe. Spectral-driven isometry-invariant matching of 3D shapes. *International Journal of Computer Vision*, 89(2-3):248–265, 2010. DOI: 10.1007/s11263-009-0250-0. 88

[RS00] S. T. Roweis and L. K. Saul. Nonlinear dimensionality reduction by locally linear embedding. *Science*, 290:2323–2326, 2000. DOI: 10.1126/science.290.5500.2323. xxiii, 45

[RS13] K. Ramani and A. Sinha. Multiscale kernels using random walks. *Computer Graphics Forum*, 33(1):164–177, 2013. DOI: 10.1111/cgf.12264. xxiii, 9, 42

[Rud87] W. Rudin. *Real and Complex Analysis*, 2nd ed., International Series in Pure and Applied Mathematics. McGraw-Hill Inc., New York, 1987. 20

[Rus07]   R. M. Rustamov.   Laplace-Beltrami eigenfunctions for deformation invariant shape representation. In *Proc. of the Symposium on Geometry Processing*, pages 225–233, 2007. xxiii, 44

[Rus11a] R. M. Rustamov. Interpolated eigenfunctions for volumetric shape processing. *The Visual Computer*, 27(11):951–961, 2011. DOI: 10.1007/s00371-011-0629-0. 9, 23

[Rus11b] R. M. Rustamov.   Multiscale biharmonic kernels.   *Computer Graphics Forum*, 30(5):1521–1531, 2011. DOI: 10.1111/j.1467-8659.2011.02026.x. xxiii, 5, 9, 23, 46

[RWP06]   M. Reuter, F.-E. Wolter, and N. Peinecke. Laplace-Beltrami spectra as Shape-DNA of surfaces and solids. *Computer-aided Design*, 38(4):342–366, 2006. DOI: 10.1016/j.cad.2005.10.011. 3, 6, 8

[RWSN09] M. Reuter, F.-E. Wolter, M. E. Shenton, and M. Niethammer. Laplace-Beltrami eigenvalues and topological features of eigenfunctions for statistical shape analysis. *Computer-aided Design*, 41(10):739–755, 2009. DOI: 10.1016/j.cad.2009.02.007. 23, 71, 75

[Saa92]   Y. Saad. Analysis of some Krylov subspace approximations to the matrix exponential operator. *SIAM Journal of Numerical Analysis*, 29:209–228, 1992. DOI: 10.1137/0729014. 20, 21

[SCOIT05] O. Sorkine, D. Cohen-Or, D. Irony, and S. Toledo. Geometry-aware bases for shape approximation. *IEEE Transactions on Visualization and Computer Graphics*, 11(2):171–180, 2005. DOI: 10.1109/tvcg.2005.33. 79, 86

[SCOT03] O. Sorkine, D. Cohen-Or, and S. Toledo. High-pass quantization for mesh encoding. In *Proc. of the Symposium on Geometry Processing*, pages 42–51, 2003. 8

[SdGP+15] J. Solomon, F. de Goes, G. Peyrè, M. Cuturi, A. Butscher, A. Nguyen, T. Du, and L. Guibas. Convolutional wasserstein distances: Efficient optimal transportation on geometric domains. *ACM Transactions on Graphics*, 34(4):66:1–66:11, July 2015. DOI: 10.1145/2766963. 46

[She02]   J. R. Shewchuk. Delaunay refinement algorithms for triangular mesh generation. *Computational Geometry Theory and Applications*, 22(1-3):21–74, 2002. DOI: 10.1016/s0925-7721(01)00047-5. 3

[Sid98]   R. B. Sidje. Expokit: A software package for computing matrix exponentials. *ACM Transactions on Mathematical Software*, 24(1):130–156, March 1998. DOI: 10.1145/285861.285868. 21

[Sin64] R. Sinkhorn. A relationship between arbitrary positive matrices and doubly stochastic matrices. *The Annals of Mathematical Statistics*, 35(2):876–879, June 1964. DOI: 10.1214/aoms/1177703591. 46

[Sin06] A. Singer. From graph to manifold Laplacian: The convergence rate. *Applied and Computational Harmonic Analysis*, 21(1):128–134, 2006. DOI: 10.1016/j.acha.2006.03.004. xxiii

[SK03] A. J. Smola and R. I. Kondor. Kernels and regularization on graphs. In *Conference on Learning Theory*, pages 144–158, 2003. DOI: 10.1007/978-3-540-45167-9_12. xxiii, 45

[SKS07] A. Spira, R. Kimmel, and N. Sochen. A short-time Beltrami kernel for smoothing images and manifolds. *Transactions on Image Processing*, 16(6):1628–1636, 2007. DOI: 10.1109/tip.2007.894253. 88

[SLCO⁺04] O. Sorkine, Y. Lipman, D. Cohen-Or, M. Alexa, C. Roessl, and H.-P. Seidel. Laplacian surface editing. In *Proc. of the Symposium on Geometry Processing*, pages 179–188, 2004. DOI: 10.1145/1057432.1057456. 9

[SM00] J. Shi and J. Malik. Normalized cuts and image segmentation. *IEEE Transactions on Pattern Analysis and Machine Intelligence*, 22(8):888–905, 2000. DOI: 10.1109/34.868688. xxiii, 45

[Sog88] C. D. Sogge. Concerning the l_p norm of spectral clusters for second-order elliptic operators on compact manifolds. *Journal of Functional Analysis*, 77(1):123–138, 1988. DOI: 10.1016/0022-1236(88)90081-x. 35, 38

[SOG09] J. Sun, M. Ovsjanikov, and L. J. Guibas. A concise and provably informative multi-scale signature based on heat diffusion. *Computer Graphics Forum*, 28(5):1383–1392, 2009. DOI: 10.1111/j.1467-8659.2009.01515.x. xxiii, 11, 12, 41, 44

[Sor92] D. C. Sorensen. Implicit application of polynomial filters in a k-step Arnoldi method. *SIAM Journal of Matrix Analysis and Applications*, 13(1):357–385, 1992. DOI: 10.1137/0613025. 8, 22, 24, 57

[Sor06] O. Sorkine. Differential representations for mesh processing. *Computer Graphics Forum*, 25(4):789–807, 2006. DOI: 10.1111/j.1467-8659.2006.00999.x. xxiii, 1

[SS02] B. Schoelkopf and A. J. Smola. *Learning with Kernels*. The MIT Press, 2002. 8

[ST07] D. A. Spielman and S.-H. Teng. Spectral partitioning works: Planar graphs and finite element meshes. *Linear Algebra and its Applications*, 421:284–305, 2007. DOI: 10.1016/j.laa.2006.07.020. xxiii, 45

[TA77]   A. N. Tikhonov and V. Y. Arsenin. *Solutions of Ill-posed Problems*. W. H. Winston Washington, D.C., 1977. 79

[Tau95]   G. Taubin. A signal processing approach to fair surface design. In *ACM Siggraph*, pages 351–358, 1995. DOI: 10.1145/218380.218473. 8, 42, 78, 79

[Tau99]   G. Taubin. 3D geometry compression and progressive transmission. In *Eurographics Tutorials*, 1999. xxiii, 1

[TGP15]   J. Tierny, D. Günther, and V. Pascucci. *Optimal General Simplification of Scalar Fields on Surfaces*, pages 57–71. Springer Berlin Heidelberg, Berlin, Heidelberg, 2015. DOI: 10.1007/978-3-662-44900-4_4. 78

[TLHD03]   Y. Tong, S. Lombeyda, A. N. Hirani, and M. Desbrun. Discrete multiscale vector field decomposition. *ACM Transactions on Graphics*, 22(3):445–452, 2003. DOI: 10.1145/882262.882290. 4, 23

[TP12]   J. Tierny and V. Pascucci. Generalized topological simplification of scalar fields on surfaces. *IEEE Transactions on Visualization and Computer Graphics*, 18(12):2005–2013, 2012. DOI: 10.1109/tvcg.2012.228. 78

[TSL00]   J. B. Tenenbaum, V. Silva, and J. C. Langford. A global geometric framework for nonlinear dimensionality reduction. *Science*, 290(5500):2319–2323, 2000. DOI: 10.1126/science.290.5500.2319. xxiii, 45

[TWBO02]   T. Tasdizen, R. Whitaker, P. Burchard, and S. Osher. Geometric surface smoothing via anisotropic diffusion of normals. In *Proc. of the Conference on Visualization*, pages 125–132, 2002. DOI: 10.1109/visual.2002.1183766. 88

[Ume88]   S. Umeyama. An eigendecomposition approach to weighted graph matching problems. *IEEE Transactions on Pattern Analysis and Machine Intelligence*, 10(5):695–703, 1988. DOI: 10.1109/34.6778. 8

[Var67]   S. R. S. Varadhan. On the behavior of the fundamental solution of the heat equation with variable coefficients. *Communications on Pure and Applied Mathematics*, 20(2):431–455, 1967. DOI: 10.1002/cpa.3160200210. 12, 46

[Var90]   R. S. Varga. *Scientific Computation on Mathematical Problems and Conjectures*. SIAM, CBMS-NSF regional conference series in applied mathematics, 1990. DOI: 10.1137/1.9781611970111. 51

[VBCG10]   A. Vaxman, M. Ben-Chen, and C. Gotsman. A multi-resolution approach to heat kernels on discrete surfaces. *ACM Transactions on Graphics*, 29(4):1–10, 2010. DOI: 10.1145/1833351.1778858. 19, 21, 66

[VL79] C. Van Loan. A note on the evaluation of matrix polynomials. *IEEE Transactions on Automatic Control*, 24(2):320–321, April 1979. DOI: 10.1109/tac.1979.1102005. 57

[VL08] B. Vallet and B. Lèvy. Spectral geometry processing with manifold harmonics. *Computer Graphics Forum*, 27(2):251–260, 2008. DOI: 10.1111/j.1467-8659.2008.01122.x. 3, 8, 17, 24, 56

[Wah90] G. Wahba. *Spline Models for Observational Data*, volume 59. SIAM, Philadelphia, 1990. DOI: 10.1137/1.9781611970128. 75, 84

[Wit83] A. P. Witkin. Scale-space filtering. In *Proc. of the International Joint Conference on Artificial Intelligence*, pages 1019–1022, 1983. DOI: 10.1016/b978-0-08-051581-6.50036-2. 88

[WMKG07] M. Wardetzky, S. Mathur, F. Kälberer, and E. Grinspun. Discrete Laplace operators: No free lunch. In *Proc. of the Symposium on Geometry Processing*, pages 33–37, 2007. DOI: 10.1145/1508044.1508063. 3

[WPG12] O. Weber, R. Poranne, and C. Gotsman. Biharmonic coordinates. *Computer Graphics Forum*, 31(8):2409–2422, December 2012. DOI: 10.1111/j.1467-8659.2012.03130.x. 5

[WZS+13] G. Wang, X. Zhang, Q. Su, J. Chen, L. Wang, Y Ma, Q. Liu, L. Xu, J. Shi, and Y. Wang. A heat kernel based cortical thickness estimation algorithm. In *MBIA*, volume 8159 of *Lecture Notes in Computer Science*, pages 233–245, 2013. DOI: 10.1007/978-3-319-02126-3_23. 88

[XHW10] B. Xiao, E. R. Hancock, and R. C. Wilsonb. Geometric characterization and clustering of graphs using heat kernel embeddings. *Image and Vision Computing*, 28(6):1003–1021, 2010. DOI: 10.1016/j.imavis.2009.05.011. xxiii, 8, 45

[Xu07] G. Xu. Discrete Laplace-Beltrami operators and their convergence. *Computer Aided Geometric Design*, 8(21):398–407, 2007. DOI: 10.1016/j.cagd.2004.07.007. 9

[XYGL13] H. Xu, W. Yu, S. Gu, and X. Li. Biharmonic volumetric mapping using fundamental solutions. *IEEE Transactions on Visualization and Computer Graphics*, 19(5):787–798, May 2013. DOI: 10.1109/tvcg.2012.173. 5

[YLL12] W. Yu, M. Li, and X. Li. Fragmented skull modeling using heat kernels. *Graphical Models*, 74(4):140–151, 2012. {GMP2012}. DOI: 10.1016/j.gmod.2012.03.011. 8

[ZF03] H. Zhang and E. Fiume. Butterworth filtering and implicit fairing of irregular meshes. In *Proc. of the Pacific Conference on Computer Graphics and Applications*, page 502, 2003. DOI: 10.1109/pccga.2003.1238303. 8, 42

[ZGL03] X. Zhu, Z. Ghahramani, and J. Lafferty. Semi-supervised learning using gaussian fields and harmonic functions. In *International Conference on Machine Learning*, pages 912–919, 2003. xxiii, 45

[ZGLG12] W. Zeng, R. Guo, F. L., and X. Gu. Discrete heat kernel determines discrete Riemannian metric. *Graphical Models*, 74(4):121–129, 2012. DOI: 10.1016/j.gmod.2012.03.009. 6, 12, 16

[ZH08] F. Zhang and E. R. Hancock. Graph spectral image smoothing using the heat kernel. *Pattern Recognition*, 41(11):3328–3342, 2008. DOI: 10.1016/j.patcog.2008.05.007. xxiii, 20

[ZKK02] G. Zigelman, R. Kimmel, and N. Kiryati. Texture mapping using surface flattening via multidimensional scaling. *IEEE Transactions on Visualization and Computer Graphics*, 8(2):198–207, 2002. DOI: 10.1109/2945.998671. 8

[ZL05] H. Zhang and R. Liu. Mesh segmentation via recursive and visually salient spectral cuts. In *Proc. of Vision, Modeling, and Visualization*, pages 429–436, 2005. 8

[ZSGS04] K. Zhou, J. Synder, B. Guo, and H.-Y. Shum. Iso-charts: Stretch-driven mesh parameterization using spectral analysis. In *Proc. of the Symposium on Geometry processing*, pages 45–54, 2004. DOI: 10.1145/1057432.1057439. 8

[ZTZX13] Y. Zheng, C.-L. Tai, E. Zhang, and P. Xu. Pairwise harmonics for shape analysis. *IEEE Transactions on Visualization and Computer Graphics*, 19(7):1172–1184, July 2013. DOI: 10.1109/tvcg.2012.309. 5

[ZvKD07] H. Zhang, O. van Kaick, and R. Dyer. Spectral methods for mesh processing and analysis. In *Eurographics State-of-the-art Report*, pages 1–22, 2007. DOI: 10.2312/egst.20071052. xxiii, 1

[ZZG+15] M. Zhang, W. Zeng, R. Guo, F. Luo, and X. D. Gu. Survey on discrete surface Ricci flow. *Journal of Computational Science and Technology*, 30(3):598–613, 2015. DOI: 10.1007/s11390-015-1548-8. 15

Author's Biography

GIUSEPPE PATANÈ

Giuseppe Patanè is a researcher at CNR-IMATI (2006–today), the Institute for Applied Mathematics and Information Technologies at the Italian National Research Council. Since 2001, his research activities have been focused on the definition of paradigms and algorithms for modeling and analyzing digital shapes and multidimensional data. He received a Ph.D. in Mathematics and Applications from the University of Genova (2005) and a Post-Lauream Master's degree in Applications of Mathematics to Industry from the F. Severi National Institute for Advanced Mathematics, Department of Mathematics and Applications at the University of Milan (2000).